Electronics texts for engineers and scientists

Editors

H. Ahmed
Reader in Microelectronics, Cavendish Laboratory,
University of Cambridge

P. J. Spreadbury
Lecturer in Engineering, University of Cambridge

Analogue and digital electronics for engineers

Analogue and digital electronics for engineers

AN INTRODUCTION

H. AHMED
Reader in Microelectronics, Cavendish Laboratory,
University of Cambridge

P. J. SPREADBURY
Lecturer in Engineering, University of Cambridge

THE SECOND EDITION OF 'ELECTRONICS FOR ENGINEERS'

CAMBRIDGE
UNIVERSITY PRESS

Published by the Press Syndicate of the University of Cambridge
The Pitt Building, Trumpington Street, Cambridge CB2 1RP
40 West 20th Street, New York, NY 10011–4211, USA
10 Stamford Road, Oakleigh, Victoria 3166, Australia

First published 1973
Reprinted 1974 1977 1978 1981 1982
Second edition 1984
Reprinted 1987 1989 1992

Printed in Great Britain at the University Press, Cambridge

Library of Congress Catalogue card number: 84-7700

British Library cataloguing in publication data

Ahmed, H.
    Analogue and digital electronics for
    engineers.—2nd ed.—(Electronics texts
    for engineers and scientists)
    1. Electronic circuits
    I. Title   II. Spreadbury, P. J.
    III. Electronics for engineers   IV. Series
    621.3815′3   TK7867

ISBN 0 521 26463 4 hardback
ISBN 0 521 31910 2 paperback

Contents

Contents

† Sections so marked may be omitted on first reading.

Contents

Contents

Contents

Preface to the second edition

The original aims of the first edition of this book entitled *Electronics for engineers* have been preserved in this new edition now called *Analogue and digital electronics for engineers*. The book remains an introductory teaching book primarily for first and second year students. It should also be of help to practising engineers and scientists wishing to know more of the fundamentals of analogue and digital chips before using them in the enormous variety of applications of electronics in modern life.

The first edition was written just after the syllabus at Cambridge University was revised and modernised and by coincidence, eleven years after the first edition, this new edition is being written just as the engineering course is again being restructured. This edition emphasises both digital and analogue electronics right from the start. It also brings the opportunity to include the book within the Cambridge University Press series 'Electronics Texts for Engineers and Scientists'.

The approach in this book is again to concentrate on the basic principles so that applications and more advanced work may be soundly based. On the other hand attempts are made throughout the text to ensure that there is enough practical and applied engineering to avoid a dry and unattractive presentation. Two entirely new chapters have been added on digital electronics and in several sections the text has been revised and supplemented to improve it or to bring it up-to-date. Chapter 7 of the first edition has been deleted to make way for new material but several of its most important topics have been included elsewhere in the text. The appendix on semiconductor principles and sections on the fabrication of integrated circuits have been deleted since this information is now usually available in other electronics courses taken by students.

Finally we are pleased to acknowledge the help of Mrs Joan Woolley with the typing of this new edition, and of Miss Jo Spreadbury with checking the manuscript.

H. A.
P. J. S.

Cambridge
January 1984

Preface to the first edition

This book is an introduction to electronic circuits for first and second year engineering students at universities, polytechnics and colleges of technology. For specialist electrical engineering students the level of presentation is designed to be suitable for a first year course. For those not specialising in electronics it should provide a complete course in electronic circuits for their first degree. Some parts of the book, such as the chapters on integrated circuit operational amplifiers and advanced circuits, should also be of use to practising engineers and scientists.

A pre-requisite to understanding this book is a course of lectures on passive circuit analysis which should include DC and AC circuits, use of mesh and nodal analysis, the superposition theorem and the propositions of Kirchhoff, Thévenin and Norton. An appendix has been included to explain the fundamental concepts of semiconductors which are needed to understand the operation of active semiconductor devices.

The approach used in this book follows our own experience in teaching a course on electronic circuits to first year engineering students at Cambridge University. The course was re-structured at the start of a new engineering degree and using the results of three years of development and experience, this book has been written. A special feature is that, in chapter 1, a discussion of the general principles of signal handling in electronic circuits, such as gain, input and output impedance, frequency response and coupling of networks, precedes the descriptions of circuits using active devices. Since these topics recur frequently throughout the whole subject of electronics they can subsequently be referred to with very little repetition of ideas. Furthermore whole amplifiers can now be purchased and many engineers only need to know how to instal them correctly into a system.

The principles of negative and positive feedback are explained in chapters 5 and 6; the treatment explores the ways in which negative feedback improves an amplifier's performance; and in oscillators the separate requirements are described that have to be met for good amplitude and frequency stability. These principles together with those of basic amplifiers described in chapter 1 are held by us to be fundamental and ageless, and should provide the core of an electronics course.

Chapters 2, 3 and 4 on f.e.t.s, bipolar transistors and integrated circuits respectively, describe how to realise amplifier blocks to fit into any system. A full treatment has been given to integrated circuit operational amplifiers. Technology may change the active devices that are available in the future and the relative importance of f.e.t.s, MOSTS and bipolar transistors could also change. These chapters may be read in any order. Chapter 7 describes more advanced circuits and applications which will be of interest to the professional engineer and to the student specialising in electronics.

Certain sections of the book have been marked with a †. These are considered to be too advanced for some first year students and may be omitted on first reading. A number of topics introduced in this book cannot be pursued fully in an introductory text and the student is encouraged to follow these in one of the more comprehensive textbooks listed in appendix C. Worked examples are included in every chapter and there are also problems from the examination papers of several universities. We have not graded these problems; instead we have included multiple-choice tests after chapters 2 4, and 6 which can be attempted by the student to check his progress and to re-assure himself that he has followed the main points as he works through the book. Answers to these tests and to the problems are included.

The book has been read in draft form by a number of students and, as a result of their comments, some sections have been improved and expanded. The help of Steven Beaumont, Malcolm Bentley and Rajan Suri is acknowledged. We have included questions from the examination papers of several universities and their permission is gratefully acknowledged.

H.A.

P.J.S.

Cambridge
November 1972

1

Principles of amplifiers

1.1 Introduction

The amplifier is a basic building block of electronic systems. The contents of the block may change over the years but we will always need to know how one amplifier will load another when they are connected in series (cascade). Also we will need to know how an amplifier will be affected by the capacitance of the wires bringing its input to it and taking its output from it.

Consider the amplifier shown in fig. 1.1. We will assume that the input X_i is related to the output X_o by a constant. The stage is said to have a gain, A, given by:

$$A = \frac{\text{amplifier output}}{\text{amplifier input}} = \frac{X_o}{X_i}.$$

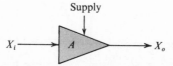

Fig. 1.1. Amplifier symbol.

Note that, in fig. 1.2, as the input X_i is increased, there will come a time when X_o cannot rise any more due to limitations of the supply. Thus every amplifier will become non-linear for very large output demands, as shown in fig. 1.2. Also all amplifiers will be non-linear to some extent, even for small signals: i.e. the ratios X_{o1}/X_{i1} and X_{o2}/X_{i2} may be different. However there will be a restricted working range of the amplifier where the ratios are nearly constant, say within a few per cent of each other. In this chapter, we consider amplifiers working in their linear range and we take A to be a constant.

Having said that the gain of an amplifier is $A = X_o/X_i$, we can write $X_o = AX_i$. If this is so, what is the output in the arrangement in fig. 1.3? Or what is the output of the mechanical system with levers shown in

1

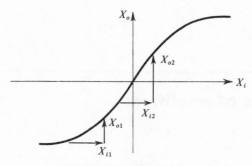

Fig. 1.2. Linearity of gain.

fig. 1.4? Is the output always $16x$ for an input x? It is clear that the output will be $16x$ sometimes; but if the output point is not entirely free to move, i.e. the output could be compressing a spring, then the beams may bend and some deflection less than 16 times the input will result. The lever example shows a realisation of $A = -4$ on no load.

Fig. 1.3. Two amplifiers in cascade.

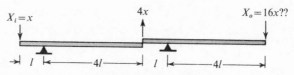

Fig. 1.4. Lever system.

Returning to the amplifiers, let us assume that they are voltage amplifiers each of gain $= -4$. That is, for a unit voltage *rise* at the input, there will be a 4 unit voltage *fall* at the output. It is again clear that the output could be $16v$ for an input v, but it may be less. This will be caused by the second amplifier drawing current by having a path of impedance Z_i at its input terminal. If the first amplifier has an impedance Z_o in its output path then a voltage drop will occur across Z_o and something less than $-4v$ will be the input of the second amplifier of fig. 1.5.

Thus we are interested in the 'coupling' between stages. Every amplifier has these input and output impedances. Can we say what are desirable values for best voltage, or current or power coupling? These impedances may be reactive; for instance, Z_i may be 10^5 Ω in parallel with a capacitor

2

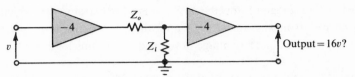

Fig. 1.5. Amplifiers showing coupling circuit.

of 10^{-10} F. We wish to know what such a value for Z_i will do to the coupling between amplifiers or between an amplifier and some signal source.

1.2 Coupling between voltage amplifiers

The circuit of fig. 1.6 contains the following parts:

(*a*) is a source of voltage v_1 and internal impedance Z_1. We know that the source may be a circuit including many devices, but Thévenin's theorem says that this can be reduced to a single voltage generator and a series impedance. We could have reduced the source to the Norton form of a single current generator and a parallel impedance; identical answers would be obtained, but less easily.

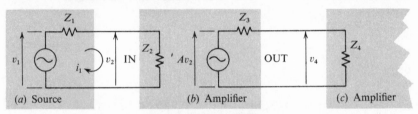

Fig. 1.6. Voltage coupling.

(*b*) is an amplifier whose input draws some current. One side of the input may be ground, or may be a power supply voltage rail which would be in common with one side of the amplifier output but, in this general case, this is not assumed. The two terminals are shown at which the input voltage v_2 may be developed and the impedance Z_2 across these terminals is the input impedance of the amplifier.

Again the amplifier may contain many components but the Thévenin representation is used whereby the output circuit is reduced to one voltage generator Av_2 and one series impedance Z_3. If no load current flows and so no voltage drop occurs in Z_3, Av_2 will be the output for an input v_2. So A is the no-load voltage gain of the amplifier and Z_3 is the amplifier's output impedance.

3

(*c*) is either a load of impedance Z_4, or is a further amplifier which may draw current from the amplifier (*b*) and which we represent as having an input impedance Z_4. The voltage developed across this load is v_4.

Considering the input circuit of the first amplifier, we wish to know what voltage v_2 is developed there compared to v_1 which is the voltage that would be available from the source if no current was drawn from it.

Kirchhoff's laws allow us to solve circuit problems in two ways. The first way is to label the voltages appearing between different points of a circuit and some reference point. We can then write the currents in each path of the circuit to be equal to the voltage difference between the ends of the path divided by the impedance of the path. Lastly we write down that the currents into any point of a circuit must equal the currents out of that point. Thus we end with as many equations as points and we can solve these equations. The second way is to mark in the unknown circulating currents in each loop and write down that the voltage drops round a loop must equal the source voltages.

In our circuit the voltages are already labelled, and at the upper input terminal of the amplifier, the current out of the source must equal that flowing into the amplifier (we have shown no other paths) thus:

$$\frac{v_1 - v_2}{Z_1} = \frac{v_2}{Z_2}$$

or

$$v_1 Z_2 = v_2 Z_1 + v_2 Z_2,$$

so

$$v_2 = v_1 \left(\frac{Z_2}{Z_1 + Z_2} \right). \tag{1.1}$$

Note that this could have been obtained by putting in i_1 as an unknown circulating current and from

$$v_1 = i_1 Z_1 + i_1 Z_2 \quad \text{and} \quad v_2 = i_1 Z_2$$

the same relation for v_2 and v_1 results.

Equation (1.1) is sometimes called the potential divider expression for the circuit of fig. 1.7.

The output voltage is a fraction $Z_2/(Z_1 + Z_2)$ of the input voltage; but note that this is only true when no other current is drawn from the

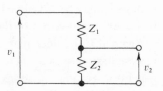

Fig. 1.7. Potentiometer.

point joining Z_1 and Z_2. This equation is easy to remember and can be used

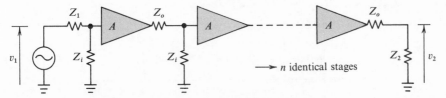

Fig. 1.8. Multistage voltage amplifier.

if all the circuits across which v_2 is developed are combined to give one effective impedance Z_2 which is used in the equation.

Under some circumstances, we may want perfect voltage coupling or $v_2 \to v_1$. From (1.1) this is achieved if

$$\frac{Z_2}{Z_1+Z_2} \to 1 \quad \text{or} \quad Z_1 \ll Z_2.$$

Note that we do not want Z_1 to be 0 but merely very much smaller than Z_2, the input resistance of the amplifier. A factor making Z_1 from 20 to 100 times smaller than Z_2 is suitable for most engineering purposes although for high quality instrumentation and computing, a factor making Z_1 from 10^3 to 10^6 smaller may be appropriate.

Referring again to the circuit, fig. 1.6, we can write an equation similar to (1.1) for the second part of the circuit:

$$v_4 = A v_2 \left(\frac{Z_4}{Z_3+Z_4}\right) = A \left(\frac{Z_2}{Z_1+Z_2}\right) \left(\frac{Z_4}{Z_3+Z_4}\right) v_1;$$

or for the multistage amplifier with n identical stages shown in fig. 1.8, we may write generally:

$$v_2 = A^n \left(\frac{Z_i}{Z_i+Z_1}\right) \left(\frac{Z_i}{Z_i+Z_o}\right)^{n-1} \left(\frac{Z_2}{Z_o+Z_2}\right) v_1, \tag{1.2}$$

where Z_i is the input impedance and Z_o is the output impedance of each amplifier.

Thus as a general method of working out the overall relation, we have separated the 'attenuation' (which is a gain of less than unity) of each coupling section between amplifier blocks from the gain of the amplifier blocks.

1.3 Worked example

A sine wave generator whose internal source is represented by a phasor of voltage $10\angle -10°$ at a certain instant has an output impedance $600-j100\ \Omega$. It is connected to a circuit of impedance $4000-j700\ \Omega$; what input will be developed there?

5

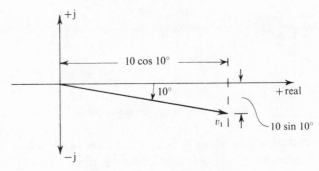

Fig. 1.9. Phasor diagram of v_1.

These data refer to a circuit similar to the left-hand part of fig. 1.6 where:

$v_1 = 10 \angle -10°$ or $10\cos 10° - j10\sin 10°$ volts (see fig. 1.9),

$Z_1 = 600 - j100\,\Omega,$

$Z_2 = 4000 - j700\,\Omega.$

Equation (1.1) gives:

$$v_2 = \frac{v_1 Z_2}{Z_1 + Z_2} = 10 \angle -10° \frac{4000 - j700}{4600 - j800}$$

$$= 10 \angle -10° \frac{4061 \angle -9°56'}{4669 \angle -9°51'}$$

$$= \frac{10 \times 4061}{4669} \angle \{-10° + (-9°\,56') - (-9°\,51')\}$$

$$= 8.70 \angle -10°\,5' \text{ volts.}$$

This is 87 per cent of the unloaded output of the generator and lags it by $5'$.

This arithmetic, which consists of taking the *product* and *quotient* of complex quantities, is more easily handled in polar form. Each step can be checked roughly and numerical mistakes are less likely to creep in, i.e. one would expect an impedance to 4000 Ω resistance and 700 Ω capacitive reactance to have a magnitude of a little over 4000 Ω, and so on.

1.4 Logarithmic expression for gain: the decibel (dB)

The logarithmic expressions for gain have several uses. To work out the gain of a multistage amplifier from an expression such as (1.2), one just *adds* the logarithmic gain for each coupling and for each amplifier to get the overall gain of many stages in cascade.

In fig. 1.8, we considered how the output voltage, v_2, was related to the input voltage v_1 and an expression of the form, $v_2 = $ (voltage gain)v_1 was obtained for an amplifier. If we know what resistances the input and output voltages appear across, say R_1 and R_2 respectively, then we can write that the input power, $p_1 = v_1^2/R_1$. Similarly the output power, $p_2 = v_2^2/R_2$, where v_1 and v_2 are the RMS voltages. Then a figure for power gain could be obtained from

$$p_2 = \text{(power gain)}p_1.$$

An alternative to the dimensionless figure for power gain is that expressed in decibels (dB) and it is defined by:

$$\text{Power gain (dB)} = 10\log_{10}\frac{p_2}{p_1}. \tag{1.3}$$

Thus an amplifier giving 4 watts output for 4mW input will have a power gain of 30 dB. If we write the powers in terms of voltages we get

$$\text{Power gain (dB)} = 10\log_{10}\frac{v_2^2/R_2}{v_1^2/R_1}$$

$$= 20\log_{10}\frac{v_2}{v_1} + 10\log_{10}\frac{R_1}{R_2}. \tag{1.4}$$

In any circuits where $R_2 = R_1$, then and *only* then:

$$\text{Power gain (dB)} = 20\log_{10}\frac{v_2}{v_1}$$

$$= 20\log_{10}\text{ (voltage gain)}. \tag{1.5}$$

Many communication circuits are designed with standard source and load resistors of 600 Ω and transmission lines of 50 and 75 Ω. In these systems, (1.5) expressing the *voltage gain* in dB is often used.

The decibel unit of power gain is very useful for two reasons:

(*a*) The response curves of many circuits have simple forms when the gain in decibels is plotted against the frequency on a logarithmic scale.

(*b*) When power (or voltage or current) amplifiers are followed by further circuits with gain ($+$dB) or attenuation ($-$dB), then the overall gain is obtained algebraically by *summing* the gains and attenuations of all parts in a path between the input and output.

Thus if our amplifier of input 4 mW and output 4 W is followed by a passive circuit of power gain 0.5 (or attenuation of 2) and this is followed by a further amplifier of power gain 400, then in terms of decibels:

$$\text{Power gain of 0.5} = 10\log_{10}0.5 = -10\log_{10}2 = -3 \text{ dB.}$$

$$\text{Power gain of 400} = 10\log_{10}400 = 10 \times 2.6 = +26 \text{ dB,}$$

and therefore the overall gain $= 30 - 3 + 26 = +53$ dB (or 200000 times).

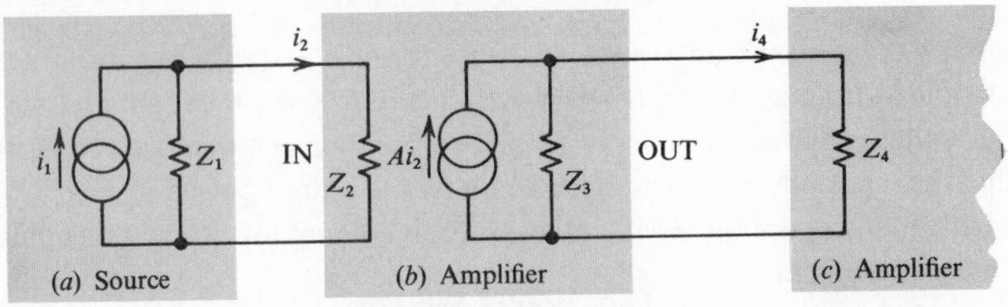

Fig. 1.10. Current coupling.

Readers will see some amplifier data sheets with the gain quoted in decibels *without* the output load being specified. These data should be treated with caution.

1.5 Coupling between current amplifiers

The circuit of fig. 1.10 contains the following parts.

(*a*) is a source of current i_1 and internal impedance Z_1. Norton's theorem allows the source, whatever its internal complexity, to be reduced to a single current generator and parallel impedance Z_1. The conduction path Z_1 means that the source can waste some of its current internally which is the dual of a voltage source dropping some of its voltage internally.

(*b*) is an amplifier whose input impedance is Z_2: thus to drive a current i_2 into the input terminals, some voltage must exist there. The amplifier is reduced to an equivalent circuit of a single generator Ai_2 and a parallel impedance Z_3 at the output terminals. Here A is the current gain of the amplifier when it is loaded by a short circuit and Z_3 is its output impedance.

In the ideal current amplifier, $Z_2 = 0$. It may appear unreal to terminate any source by a short circuit, but certain devices are good current amplifiers and we shall show that, for best current coupling, they should be followed by low impedance stages.

(*c*) is a further amplifier which has an input impedance Z_4 or it is a load.

Considering the input circuit of the first amplifier, we wish to know what current, i_2, flows into the first amplifier compared with i_1, which is apparently available from the source. Kirchhoff's current law allows us to write that the current going down Z_1 is $i_1 - i_2$, whence we can write expressions for the voltage at the amplifier input as

$$Z_1(i_1 - i_2) \quad \text{or} \quad i_2 Z_2.$$

Since these are identical,
$$Z_1(i_1 - i_2) = i_2 Z_2,$$

so
$$i_2 = i_1 \left(\frac{Z_1}{Z_1 + Z_2}\right). \tag{1.6}$$

Note that this is the dual of the voltage coupling expression. Now it is the load impedance Z_2 which should be much smaller than the source impedance to give good current coupling, i.e. when,

$$Z_2 \ll Z_1, \qquad (Z_1 + Z_2) \approx Z_1,$$

so
$$i_2 \approx i_1.$$

Consider next the current i_4 coupled from the amplifier output to the next stage input in fig. 1.10; here

$$i_4 = A i_2 \left(\frac{Z_3}{Z_3 + Z_4}\right) = A \left(\frac{Z_1}{Z_1 + Z_2}\right) \left(\frac{Z_3}{Z_3 + Z_4}\right) i_1.$$

So the overall gain, i_4/i_1, is the product of the gain of an amplifier and the efficiency of its input and output coupling; namely $Z_1/(Z_1 + Z_2)$ and $Z_3/(Z_3 + Z_4)$ respectively. If these figures are known or calculated in decibels, the terms are added rather than multiplied.

It is interesting that bipolar transistors whose input resistances are usually much smaller than their output resistances give good current coupling between each stage when they are connected in cascade. (Typical figures for a small transistor are an input resistance of 1 kΩ and an output resistance of 30 kΩ or, for a power transistor, 10 Ω and 200 Ω respectively.)

1.6 Loading of a source for maximum power output

The circuit of fig. 1.11 shows:

(*a*) a source whose voltage on no load is v_1 and whose internal impedance Z_1, has resistive and reactive components, R_1 and X_1.

(*b*) a load whose impedance, Z_2, we wish to determine to get maximum power output from the source. The load is considered generally to be made up of a resistive part R_2 in series with a reactive part of impedance X_2.

We can write the value of the current into the load, i_2, as:

$$i_2 = \frac{\text{e.m.f.}}{\text{impedance}} = \frac{v_1}{Z_1 + Z_2} = \frac{v_1}{R_1 + R_2 + j(X_1 + X_2)}. \tag{1.7}$$

The expressions for power in any load are $|v_2| \, |i_2| \cos \phi$ or $|i_2|^2 R_2$. Both

9

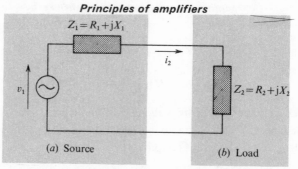

Fig. 1.11. Power coupling.

will give the same result. Here we do not know the load voltage, v_2, but we do know the load current i_2 and the resistive part of the load R_2 through which the current flows. Hence the power into the load, p_2 is given by:

$$p_2 = |i_2|^2 R_2 = \left| \frac{v_1}{R_1 + R_2 + j(X_1 + X_2)} \right|^2 R_2$$

which cannot exceed

$$\frac{v_1^2 R_2}{(R_1 + R_2)^2}. \tag{1.8}$$

This is because the complex denominator can be made a minimum if $X_1 = -X_2$ and this will make the power a maximum. Thus the first condition for getting maximum power output is that the load reactance, X_2, should be the conjugate of the source internal reactance, X_1; i.e. if one is inductive the other should be capacitative and vice versa.

Differentiating the expression for power, p_2, with respect to R_2 gives:

$$\frac{dp_2}{dR_2} = \frac{(R_1 + R_2)^2 v_1^2 - 2v_1^2 R_2 (R_1 + R_2)}{(R_1 + R_2)^4}$$

$= 0$ for a maximum or minimum. This is given by setting the numerator to zero,

$$0 = (R_1 + R_2) - 2R_2,$$

hence

$$R_1 = R_2. \tag{1.9}$$

This is a well known result that the resistive part of the load impedance must be equal to the resistive part of the source impedance for maximum power output. Substituting the value $R_2 = R_1$ into the expression for power p_2, (1.8), gives:

$$p_2 \, (\text{maximum}) = \frac{1}{4} \frac{v_1^2}{R_1}.$$

Note that v_1^2 / R_1 is the power that could be dissipated internally in the source if the output terminals were loaded only by a reactance X_2 equal to $-X_1$. Thus the maximum power output is a quarter of that that could be dissipated internally in the source.

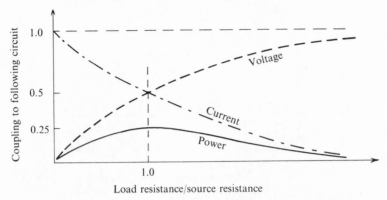

Fig. 1.12. Variation of coupling with the ratio of load to source resistance.

Power matching is used mainly in three situations:

(*a*) Where the signal levels are very small so any power lost gives a much worse signal to noise ratio: for example, it is used in all aerial to receiver connections in television, radio and radar equipment.

(*b*) Where the signal at high frequency is connected through lines of appreciable self capacity and self inductance to a load. Then it is possible to get large standing waves due to reflections from the load which can make the source to load power transfer low.

(*c*) Where signals are very large, say at the output stage of a transmitter, and where the maximum efficiency is desirable on economic grounds.

Fig. 1.12, summarises how the ratio of load to source resistance influences the efficiency of voltage, current and power coupling between circuits.

1.7 Frequency characteristics of coupling circuits and amplifiers

So far, the gain and other properties of the amplifier blocks and the coupling between the blocks have not been related in any way to frequency. There will always be some high frequency at which the gain of any amplifier is less than its gain at low frequency. The effects are akin to mechanical inertia; on requiring an output, it takes a finite time for current flowing at an input terminal of a device to pass through it and reach its output – this is called transit time and it is naturally very short if the device is small. Also, it takes time for the voltage to build up in an output circuit after current has started to flow into it, due to its capacitance. Thus we can show some features of an amplifier's response by showing the output of an amplifier for a step input, fig. 1.13.

11

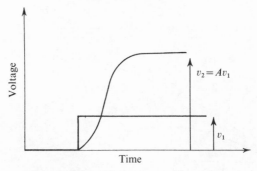

Fig. 1.13. Amplifier output for step input.

To a rough approximation, the output may rise exponentially; the time constant of the amplifier can be obtained from the time taken for the output to get to 63 per cent or $1 - 1/e$ of the state to which it eventually settles. If this time is τ_2, then we have approximated the output as a function of time to $v_2(t) = Av_1(1 - e^{-t/\tau_2})$.

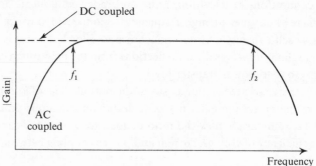

Fig. 1.14. Frequency response of a typical circuit.

An alternative representation to show the gain of an amplifier or of a coupling circuit is to plot the magnitude of its gain against frequency, as in fig. 1.14. The approximate relationship between f_2 when the gain starts dropping and τ_2 is simple and it will be developed later.

The important features of an amplifier can be seen from a plot of its |gain| against frequency. It will amplify signals at frequencies between f_1 and f_2 by very nearly the same amount. Thus a musical note with certain harmonics at the source should be almost faithfully amplified if the fundamental and the main harmonics present are in the range of frequencies between f_1 and f_2. The waveform of the electrical signal to make a television picture is very complex. It contains components between 25 Hz and 5 MHz and thus 'video' amplifiers need to have a flat response over this band.

12

The methods of Fourier series and Fourier transforms allow pulses, ramp waveforms and square waves to be broken up into a spectrum of frequencies. The circuit should have approximately the same gain and approximately the same phase shift over this spectrum for the waveform at the output to closely resemble that at the input.

We wish therefore to know which components in the coupling and amplifier circuits will give rise to the frequencies f_1 and f_2 at which the gain drops from a largely constant figure called the *midband gain*. It is possible to have coupling circuits whose gain stays constant down to zero frequency – such circuits are called direct current coupled or 'DC coupled'. With many circuits this is not necessary and they can be simplified by being coupled for alternating signals only and so are called 'AC coupled'. The difference between these two is shown at the left-hand end of the response plot, fig. 1.14.

We want to construct such a plot with the minimum of computational effort. We wish to know which components in our amplifiers and coupling circuits give rise to the drop in gain at the extremes of frequency. Then we can design circuits to handle just the frequency band required for the signals in which we may be interested.

1.8   Coupling circuits at low frequency

The following analysis explores the effect of a simple coupling circuit between a source and load. Exactly the same effect is produced by coupling circuits between stages of amplifiers and by the decoupling circuits needed by amplifiers (decoupling circuits are mentioned in the chapters on amplifier realisation with field-effect and bipolar transistors). In those cases one must first identify the resistors in series with the reactive element and the analysis then reduces to the same as that now given.

The circuit, fig. 1.15, shows:

(*a*) is a sinusoidal source of voltage of amplitude v_1 and frequency ω rad/s $= 2\pi f$ where f is in hertz (Hz). The source internal impedance Z_1 is shown to be partly resistive and partly reactive. The reactive part is due to a capacitor C_1 which may be blocking the internal supply voltages of the source from appearing at its output terminal. Alternatively it may be stopping the source circuit from upsetting the DC supply voltages of an amplifier which is forming the first stage of the load. Then C_1 may actually be within the load but it can be considered as having an effect similar to that of R_1; i.e. a voltage drop will occur across it so that part of v_1 is dropped and not usefully developed across the resistive part of the load.

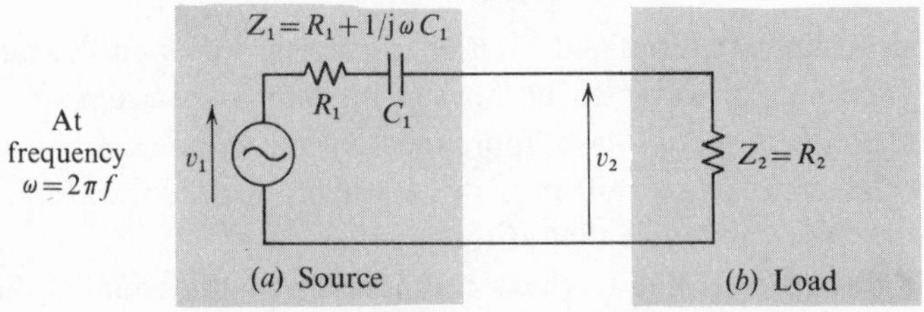

Fig. 1.15. Coupling circuit with series capacitor.

(b) is the load which is considered to be resistive only, so $Z_2 = R_2$. This is an approximation because there will be stray capacities in all circuits, but the fact is separately investigated in §1.9.

We need to obtain an expression for the voltage v_2 appearing at the load in terms of the source voltage v_1. The expression for voltage coupling gives

$$v_2 = v_1 \left(\frac{Z_2}{Z_1 + Z_2} \right) = v_1 \left(\frac{R_2}{R_1 + 1/j\omega C_1 + R_2} \right) = \frac{v_1 R_2/(R_1 + R_2)}{1 + 1/j\omega C_1(R_1 + R_2)}. \quad (1.10)$$

By studying the dimensions of each term of the final expression, it can be seen that the denominator must be dimensionless. Since ω has the units of s^{-1}, the product $C_1(R_1 + R_2)$ must have the units of seconds and is called the time constant, τ_1. (This seems odd until the units of farads × ohms are looked at more closely and are found to be seconds!)

Now we wish to find out how the expression for v_2 changes with the frequency, ω. We consider three cases.

(a) Frequency is high so ω in the denominator of the expression makes the complex part $\ll 1$.

$$\text{So gain} = \frac{v_2}{v_1} \approx \frac{R_2}{R_1 + R_2} = B, \quad \text{say}. \quad (1.11)$$

Note that this is real and not imaginary. Therefore there is no phase shift associated with the circuit gain at high frequency. This is plotted in fig. 1.17 as the region (a).

(b) At a frequency $\omega = \omega_1$, such that

$$\omega_1 C_1(R_1 + R_2) = 1 \quad \left(\text{i.e. } \omega_1 = \frac{1}{C_1(R_1 + R_2)} = \frac{1}{\tau_1} \right). \quad (1.12)$$

$$\text{Gain} = \frac{v_2}{v_1} = \frac{R_2/(R_1 + R_2)}{1 + 1/j.1} = \frac{B}{1 - j}$$

$$= \frac{B(1 + j)}{2} = 0.707B \angle + 45°.$$

Fig. 1.16 shows the conversion from cartesian to polar co-ordinates.

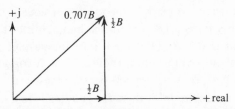

Fig. 1.16. Phasor diagram for gain at low frequency turnover.

At this particular frequency, ω_1, the gain is 70.7 per cent of its value at the higher frequency and the phase shift is such that the output *leads* the input by 45°. This is plotted on fig. 1.17 in the region (b). ω_1 is called the *turnover frequency*.

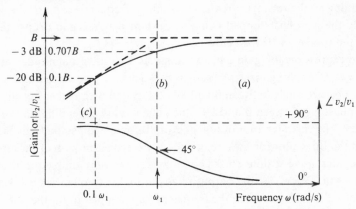

Fig. 1.17. Gain magnitude and phase angle plotted against frequency.

(c) The third region of interest is at very low frequency, well below the turnover frequency ω_1. Remember that at ω_1, the complex part of the denominator of the expression had equalled unity, so, at a much lower frequency,

$$\frac{1}{\omega C_1(R_1+R_2)} \gg 1.$$

Using

$$C_1(R_1+R_2) = \frac{1}{\omega_1},$$

then from (1.10)

$$\text{Gain} = \frac{v_2}{v_1} \approx \frac{R_2/(R_1+R_2)}{\omega_1/j\omega} = 0+jB\frac{\omega}{\omega_1} = B\frac{\omega}{\omega_1} \angle +90°. \quad (1.13)$$

The final expression has been converted from complex terms to polar co-ordinates to give a magnitude and phase angle for the gain and these are shown in region (c) of figure 1.17.

15

Thus at a frequency one tenth of the turnover frequency, $\omega = 0.1\omega_1$, the magnitude of the gain is $0.1B$, and at one hundredth of the turnover frequency, the gain is $0.01B$. Thus on a logarithmic plot of gain against frequency, the relation will be linear (note $0.1B = 20 \log_{10} B - 20$ decibels and $0.01B = 20 \log_{10} B - 40$ decibels because these are voltage ratios).

The broken lines on fig. 1.17 where the gain tends to a constant value B at frequencies above the frequency ω_1 and where the gain tends to fall linearly at frequencies below ω_1 show an approximate response curve which is called the asymptotic approximation to the frequency response. More accurately, the gain has dropped to $0.707B$ at ω_1 and the response curve really passes through this point and is an asymptote to the two broken lines which are accurate for low or high frequencies. The reason for calling ω_1 the turnover frequency or break frequency can now be seen; clearly the gain has stopped being a constant figure B and drops steeply once the frequency drops below ω_1.

The expression for gain gave an angle as well as a magnitude for the ratio v_2/v_1. Rewriting the expression as $v_2 = \text{gain} \times v_1$, we see that if the gain had a positive angle between 0 and 90° associated with it then v_2 will lead v_1 in phase by between 0 and 90°, the actual value depending on the frequency. The cartesian plot of $\log |\text{gain}|$ and $\angle \text{gain}$ against log frequency is called the Bode plot for the circuit and fully specifies the gain of the circuit.

The alternative scaling on the $|\text{gain}|$ axis shows how simple the relation becomes in decibels. If the output of Bv_1 at higher frequencies is taken as the normal level, then when the frequency has dropped to ω_1, the output is 0.707 of Bv_1.

$$\text{Voltage ratio } 0.707 = 20\log_{10}0.707$$
$$= -20\log_{10}1.414$$
$$\approx -3 \text{ decibels.}$$

When the frequency is well below ω_1, if the frequency is halved, or dropped by an octave, the gain will also be halved.

$$\text{Voltage ratio } 0.5 = 20\log_{10}0.5$$
$$= -20\log_{10}2.0 \approx -6 \text{ decibels.}$$

So the slope of the asymptote is -6 decibels/octave. It can alternatively be expressed as -20 decibels/decade.

The analysis from fig. 1.15 showed the source as a voltage generator. A similar analysis could be done if the source was a current generator, i_1, and had a parallel conductance G_1. However it is easier to use the conversion from current to voltage source shown in fig. 1.18.

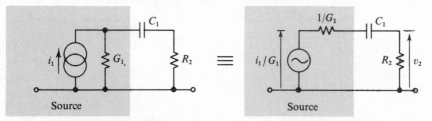

Fig. 1.18. Conversion from current to voltage controlled source.

Then we can write the output voltage, v_2, at midband directly from (1.11) as

$$v_2 = \frac{i_1}{G_1} \frac{R_2}{R_2 + 1/G_1} = i_1 \frac{R_2}{1 + G_1 R_2} \qquad (1.14)$$

and the turnover frequency, ω_1, from (1.12) as

$$\omega_1 = \frac{1}{\tau_1} = \frac{1}{C_1 \left(\dfrac{1}{G_1} + R_2 \right)}. \qquad (1.15)$$

Practically, it can be seen from (1.12) or (1.15) that the turnover frequency can be made low by choosing a large value for the coupling capacitor C_1 or a large value for either the source resistance $R_1 (= 1/G_1)$ or load resistance R_2. At the low working voltages of transistor and integrated circuits, a capacitor C_1 of value 100 μF may not be too large or expensive if an extended low frequency response is required. If this is not sufficient, then it may be possible to choose circuits, where R_1 or R_2 is made very high. In voltage amplifiers, for good coupling (see (1.1)) we required R_2 to be much greater than R_1 so it is better to concentrate on making R_2, the input resistance of the following circuit, as high as possible. This may be achieved by choosing the components in the input circuit to be as high a resistance as the makers' data will allow, or by using negative feedback (see chapter 5). Alternatively, in current amplifiers, for good coupling (see (1.6)) we require R_2 to be much *less* than R_1, so it is better to concentrate on making R_1, the output resistance of the source, high if we also want good low frequency response. Again this can be achieved by the use of negative feedback.

Lastly, if the response is required not to drop at very low frequencies, it is better to couple the source directly to the load and to dispense with the blocking capacitor, C_1. This may need extra circuit complexity (complementary transistors, Zener diodes, positive and negative supplies) but it avoids the use of bulky and probably costly capacitors. This topic is mentioned in more detail in chapter 4.

17

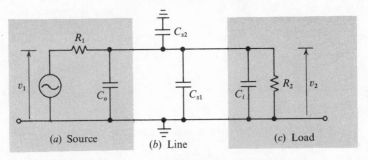

Fig. 1.19. Coupling circuit with shunting capacitors.

1.9 Coupling circuits at high frequency

The circuit of fig. 1.19 contains the following parts:

(*a*) is a voltage source, as in the previous example, but now with some effective output capacity, C_o, across its terminals. This may be the sum of the internal capacity in the output device of the source and stray capacities of wiring and components in the output circuit.

(*b*) is the line joining the source to load. This is shown with the lower wire at earth; the upper wire then has capacitance C_{s1} to the lower wire and further stray capacitance C_{s2} to all other nearby earthed objects. These capacities can be made small by spacing the wire well away from all earthed objects. However if the signal levels are low, the signal wire must be screened to prevent electrostatic pickup. So the line may be a coaxial cable and C_s may be of the order of a few tens of picofarads for each metre of length of the line joining source to load.

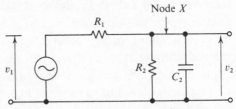

Fig. 1.20. Equivalent circuit of fig. 1.19.

(*c*) is the load which will have some capacity, C_i, between its input terminals as well as a resistive input path, R_2.

If we combine all the capacities into one component C_2 where

$$C_2 = C_o + \Sigma C_s + C_i \tag{1.16}$$

then the circuit simplifies to that shown in fig. 1.20.

18

To relate v_1 to v_2 in this circuit, we can write the current flowing to the right through R_1 as $(v_1 - v_2)/R_1$, the current down R_2 as v_2/R_2 and the current down C_2 as $v_2 j\omega C_2$. Then if no other current is drawn from node X, Kirchhoff's current law gives:

$$\frac{v_1 - v_2}{R_1} = \frac{v_2}{R_2} + v_2 j\omega C_2.$$

$$\text{Gain} = \frac{v_2}{v_1} = \frac{R_2}{R_1 + R_2 + j\omega C_2 R_1 R_2} = \frac{R_2/(R_1 + R_2)}{1 + j\omega C_2 R_1 R_2/(R_1 + R_2)}. \quad (1.17)$$

Note that each term of this expression is dimensionless; the numerator is B, the circuit gain at midband. Since ω has the units of s^{-1},

$$C_2 R_1 R_2/(R_1 + R_2)$$

has the units of seconds and is the time constant, τ_2. Note that the C–R product is the total *shunting* capacity multiplied by the *parallel value* of the two circuit resistors.

We wish to find out how the expression for gain varies with frequency, ω. Again it can be considered in three regions.

(*a*) Frequency is low, so ω in the denominator of the expression makes the complex part $\ll 1$. So

$$\text{Gain} = \frac{v_2}{v_1} \approx \frac{R_2}{R_1 + R_2} = B.$$

Note that this is real. Therefore there is no phase shift associated with the circuit gain at low frequencies. This is plotted in fig. 1.22 as region (*a*).

(*b*) At a frequency $\omega = \omega_2$, when the imaginary part of denominator = 1,

$$\omega_2 C_2 \left(\frac{R_1 R_2}{R_1 + R_2} \right) = 1 \quad \text{or} \quad \omega_2 = \frac{R_1 + R_2}{C_2(R_1 R_2)} = \frac{1}{\tau_2} \quad (1.18)$$

then the expression for gain becomes

$$\text{Gain} = \frac{v_2}{v_1} = \frac{B}{1 + j \cdot 1} = \frac{B(1-j)}{2} = 0.707B \angle -45°.$$

Fig. 1.21 shows the conversion from cartesian to polar co-ordinates.

At the particular frequency, ω_2, the upper turnover frequency, the gain drops to 70.7 per cent of its value at lower frequencies and the phase shift is such that the output lags the input by 45°. This may be seen by writing the expression for gain as $v_2 = 0.707Bv_1 \angle -45°$ which is plotted on fig. 1.22 as the region (*b*).

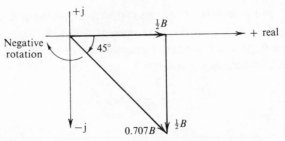

Fig. 1.21. Phasor diagram for gain at high frequency turnover.

(*c*) At a very high frequency, well above the turnover frequency, ω_2, where the complex part of the denominator of the expression equals unity,

$$j\omega C_2 R_1 R_2/(R_1+R_2) \gg 1$$

and using $\qquad C_2 R_1 R_2/(R_1+R_2) = 1/\omega_2$ in (1.17),

$$\text{Gain} \sim \frac{v_2}{v_1} = \frac{R_2/(R_1+R_2)}{j\omega/\omega_2} = B\omega_2/\omega \angle -90°. \qquad (1.19)$$

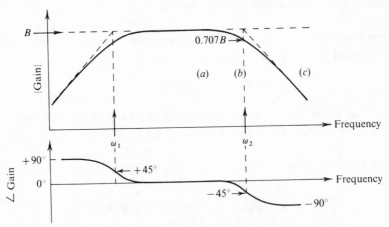

Fig. 1.22. Log gain magnitude and phase angle plotted against log frequency.

This is shown in region (*c*) of fig. 1.22. As the frequency ω becomes very much higher than ω_2, say $10\omega_2$ or $100\omega_2$, |gain| becomes $0.1B$ or $0.01B$ and the phase shift settles very close to 90° (the output lagging the input).

This plot of |gain| on a logarithmic scale (or in decibels) and $\angle$ gain versus frequency on a logarithimic scale is a Bode plot. Fig. 1.22 is for a circuit whose gain drops at high frequencies, ω_2, because of shunting capacities C_2, *and* at low frequencies, ω_1, because of coupling capacities. Provided that

ω_1 and ω_2 are several orders of magnitude apart, then we can calculate separately the effects of coupling capacities and shunting capacities. For example, in a normal amplifier the output coupling capacitor may be several μF and the stray capacities several hundred pF. Since these are a factor of 10^4 apart the two separate effects arising from them are often not both noticeable at any one frequency.

The frequency response is shown clearly in fig. 1.22. The circuit has a nearly constant gain, B, in the region between ω_1 and ω_2. At each of these particular frequencies the voltage gain drops to $0.707\,B$ (or $1/\sqrt{2}$ of normal). If we consider the output power, p_2,

$$p_2 \propto (v_2)^2$$

and so drops to half its normal value. So the frequencies ω_1 and ω_2 are called the *half power* frequencies and $\omega_2 - \omega_1$ is the *half power bandwidth*.

The conditions at the half power frequencies can be written in another way which is easier to remember. At the lower turnover frequency, ω_1, (1.12) gave:

$$\omega_1 C_1 (R_1 + R_2) = 1 \quad \text{or} \quad \frac{1}{\omega_1 C_1} = R_1 + R_2 \tag{1.20}$$

so the impedance of the *series* capacitor = *series* effect of R_1 and R_2.

At the higher turnover frequency ω_2, (1.18) gave:

$$\frac{\omega_2 C_2 R_1 R_2}{(R_1 + R_2)} = 1 \quad \text{or} \quad \frac{1}{\omega_2 C_2} = \frac{R_1 R_2}{R_1 + R_2} \tag{1.21}$$

so the impedance of the *shunting* capacities = *parallel* effect of R_1 and R_2.

Lastly, let us look at how the response of an amplifier will be affected by the coupling circuit in series with it. Fig. 1.23 shows the gain of the amplifier (a) and that of the coupling circuit (b). If the gains are in decibels, then they are added to give (c), the gain of the whole circuit. ω_2 is the turnover frequency of the coupling circuit and ω_3 is the turnover frequency of

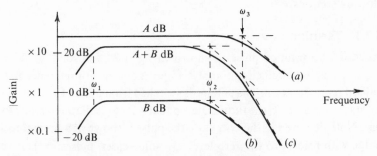

Fig. 1.23. Combination of gain plots: (a) amplifier alone, (b) coupling circuit, and (c) amplifier and coupling circuit.

21

the amplifier alone. This causes the gain of the whole circuit to drop at 6 dB/octave from ω_2 to ω_3 and then at 12 dB/octave from ω_3 upwards. Also if the gains in midband are A decibels for the amplifier and B decibels for the coupling circuit, then the amplifier and coupling circuit will have a gain of $A+B$ decibels. Fig. 1.23 shows $A+B$ just below A because the coupling circuit attenuates.

1.10 Pulses and digital signals

So far we have considered continuously varying sinusoidal signal waveforms or steady voltage levels. In *digital* electronics the signals are not continuously varying; instead, the voltage or current at the output terminals of a circuit is in the form of a pulse. It makes a rapid transition from one value or level to another and after an interval of time, usually long in comparison with the transition time, returns rapidly to its original value. For the higher level of output measured in terms of current or voltage the circuit is said to be HIGH while for the smaller value, often close to zero, the circuit is in the LOW state. The duration of the pulse in one or other of its states can vary over a very wide range of times ranging from a few nanoseconds (1 ns $= 10^{-9}$ s) to several seconds. The time taken to switch from one state to the other is not instantaneous but it is usual practice to try and make it a small fraction of the duration of the pulse and in most digital circuits and systems it is assumed that the time is negligibly short. The shape of a pulse, in terms of its amplitude plotted as a function of time, is determined by the combination of the passive components such as resistors, capacitors and inductors that make up the circuit and by the operating characteristics of the active devices such as diodes and transistors with which the pulse is generated. The main feature of pulse waveforms and the factors that govern the shapes of pulses are described in the following sections.

1.10.1 Terminology

A portion of a train of pulses is shown in fig. 1.24 and some of the terminology met in pulsed circuit analysis is introduced with reference to this idealised representation. The voltage is switched from a level V_1 which is the LOW state to a level V_2, the HIGH state, so that $V_2 - V_1$ is the *amplitude* of the pulse. Note that even in the LOW state the pulse is shown at a finite positive voltage with respect to the zero level. In some cases, pulses in their LOW state may be at the zero level but this is not a necessary condition: indeed the low level may be either a small positive or negative amount away from

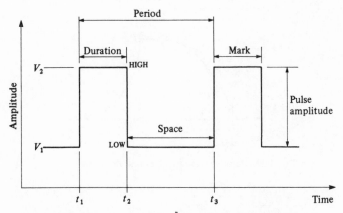

Fig. 1.24. Part of a train of pulses and the terminology used.

zero. In this ideal case the pulse amplitude is assumed to be constant over the whole of its duration in either of its states. In practice, however, there may be some small changes in amplitude while the pulse is in the HIGH or LOW state.

The time for which the pulse is at the level V_2 is the pulse duration in the HIGH state, i.e. $t_2 - t_1$. The ratio of the time for which the pulse is HIGH to the time for which it is LOW is called the mark-to-space ratio. The *duty cycle* is the time for which the pulse is HIGH in a period, i.e. $(t_2 - t_1)/(t_3 - t_1)$. The *repetition* frequency of a pulse train is the number of pulses per second or $1/(t_3 - t_1)$ Hz. For example, the pulse duration of the pulse train shown in fig. 1.25 is 1 μs, the mark-to-space ratio is $\frac{1}{9}$, the duty cycle is 1 in 10 and the pulse repetition frequency is 10^5 pulses/second. The pulse amplitude is 3 V offset at $+2$ V with respect to the zero level.

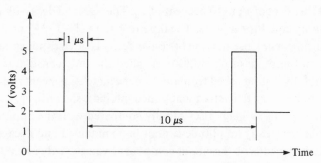

Fig. 1.25. An example of a pulse waveform.

23

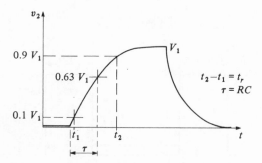

Fig. 1.26. Pulse shape with rise- and decay-times.

1.10.2 Rise-time, decay-time and time constants

We have assumed in § 1.10 that the time taken to change from V_1 to V_2 is infinitesimally small. This is not always true although it is again emphasised that it is a perfectly valid assumption in most digital circuit analysis. In some applications it is not possible to make this time a negligibly small fraction of the pulse duration. Generally, as described in § 1.7, the pulse shape is as shown in detail in fig. 1.26. The pulse voltage increases gradually from a base or zero level to a level V_1 in a time determined by circuit components. It can be seen in fig. 1.26 that the time at which the pulse reaches the voltage V_1 exactly is not precisely defined. Thus it is difficult to state accurately the total time taken to switch the voltage from zero to V_1 and it is frequently necessary to know how quickly or how slowly the pulse amplitude is changing towards V_1. To overcome this difficulty a 'rise-time' is defined as the time in which the pulse voltage rises from 10 per cent to 90 per cent of its final value: these are precise conditions. This method gives the two points shown in fig. 1.26 and the rise-time $t_r = t_2 - t_1$. In a similar manner a fall or decay-time may be defined as the time taken to fall from 90 per cent of V_1 to 10 per cent of V_1. The rise- and decay-times are not necessarily equal for a pulse. The time to rise to $(1 - 1/e)V_1$ or $0.63V_1$ is also of significance since the rate of rise or decay is often exponential and is marked on the pulse in fig. 1.26. It is called the time constant (τ) and must be carefully distinguished from the 10 per cent to 90 per cent rise-time.

Another pulse shape that is frequently encountered is shown in fig. 1.27(a). In this case the pulse voltage rises rapidly compared with the duration of the pulse but the steady amplitude cannot be maintained and decays at a rate that again depends on the circuit component values. (Other waveforms that are frequently encountered in pulse circuits are a pulse with over-

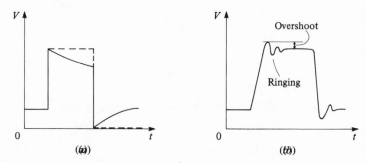

Fig. 1.27. (*a*) and (*b*) pulse waveforms.

shoot and ringing (fig. 1.27(*b*)) and the square wave with equal HIGH and LOW periods as shown in fig. 1.28(*a*), the sawtooth waveform (fig. 1.28(*b*)) and the triangular waveform (fig. 1.28(*c*)).)

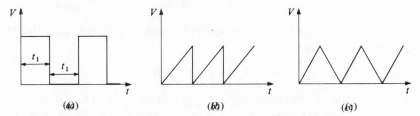

Fig. 1.28. (*a*) Square wave, (*b*) sawtooth, (*c*) triangular.

1.10.3 Resistance–capacitance coupling network

One of the circuits most often encountered in electronics is shown in fig. 1.29. The capacitance C and the resistance R may be actual components or they may arise from the wiring resistances and package capacitances or connections used in the coupling of one circuit to another. For example, in the measurement of the output signal from an amplifier, the output resistance of the amplifier, the capacitance of co-axial leads and the oscilloscope's input capacity may form just such a circuit as shown in fig. 1.29. The circuit may be regarded as a potential divider for sinusoidal signal inputs whose frequency is variable. For very high frequencies the voltage across the capacitor is very small since its reactance is small compared with the value of the resistance while at low frequencies the voltage across the capacitor is large and the converse is true. For a voltage step that forms the first part of a pulse the output may be obtained analytically after noting that the current through R must flow into C; this current is $C\,dv_2/dt$.

25

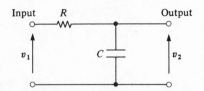

Fig. 1.29. A resistance–capacitance coupling network.

Kirchhoff's voltage law gives:

$$v_1 = RC\frac{dv_2}{dt} + v_2 = (sRC + 1)\,v_2,$$

where s is the differential operator ($\equiv D$ in some textbooks),

$$v_2 = \frac{1}{1 + sCR}v_1$$

When the input voltage changes suddenly from zero to the value V_1 the equation has a solution:

$$v_2 = V_1[1 - \exp(-t/RC)].$$

Thus the value of the output voltage v_2 is a function of time and depends on the value of the resistance–capacitance product, RC. In fig. 1.30 we show a plot of v_2 as a function of time. It is of interest to note that at $t = RC$ (the time constant, τ), $v_2 = V_1(1 - 1/e)$, or 63.2 per cent of V_1. We can also determine the 10 per cent to 90 per cent *rise-time* $(t_2 - t_1)$ in terms of RC.

If $v_2 = 0.1V_1$, $t_1 = 0.1RC$ and when $v_2 = 0.9V_1$, $t_2 = 2.3RC$. Therefore the *rise-time* $t_r = t_2 - t_1 = 2.2RC$. The time constant $\tau\,(= RC)$ of the circuit should not be confused with the 10 per cent to 90 per cent rise-time.

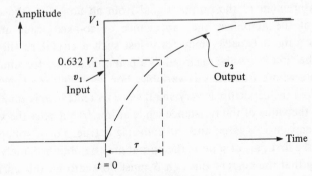

Fig. 1.30. Rise-time definitions.

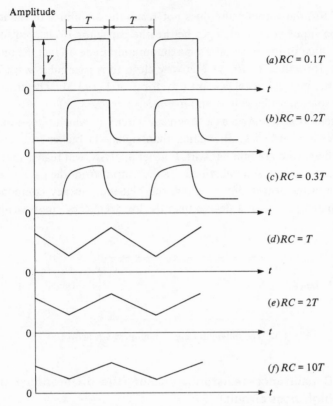

Fig. 1.31. Output waveforms for an ideal square wave applied to an *RC* network. From (*a*) to (*f*) the time constant *RC* changes from *RC* = 0.1*T* to *RC* = 10*T*.

1.10.4 Square wave applied to an *RC* network

An ideal square waveform pulse train of amplitude V and period $2T$ can be used to determine the response of the *RC* network. If it is applied to the circuit of fig. 1.29 the output waveform depends on the ratio of the time constant $\tau = RC$ and the periodic time. We assume that RC may be varied. For example, if $RC = 0.1T$, so that the rise-time is much shorter than the period, then the waveform corresponds to that shown in fig. 1.31(*a*). The circuit output, for a change of input from 0 to V_1, reaches a value, after time T, of $v_2 = V_1[1 - \exp(-10)] = 0.999\,955 V_1$. For $RC = 0.2T$ the waveform amplitude reaches $0.993 V_1$, for $RC = 0.3T$ the waveform reaches only 96.4 per cent of V_1 and for $RC = T$ the waveform shown in fig. 1.31(*d*) reaches only to 63.2 per cent of V_1. For $RC = 2T$ the waveform reaches 39 per cent of V and for $RC = 10T$ the waveform reaches to just 9.5 per cent of V_1 as shown in fig. 1.31(*f*). Thus it is seen that, for the larger

27

values of RC, the output pulse does not reach the amplitude of the input pulse. The input voltage changes before the capacitor is charged to the required value of the waveform's maximum amplitude. In fact the output waveform for values of $RC \gg T$ is very close to a triangular waveform which is the integral of the square waveform of Fig. 1.31. The circuit is therefore sometimes known as an *integrating circuit*.

The circuit is also known as a 'low-pass' circuit in which signals at low frequencies are much less attenuated than signals at higher frequencies. Some readers, who are familiar with Fourier analysis, will realise that this is borne out by the results obtained for the output from the square wave input. The waveform is one in which the higher frequency components are attenuated to a greater degree than the low frequency components.

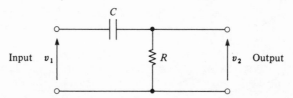

Fig. 1.32. The differentiating or high-pass CR network.

1.10.5 Capacitance–resistance circuit (the differentiating or high-pass circuit)

The circuit shown in fig. 1.32 is another resistance–capacitance network but with the two circuit components exchanged compared with the circuit of fig. 1.29. If a step input voltage is applied to this circuit the output voltage follows the pulse very closely at first but then decays while the input voltage remains constant. When the input pulse turns off, the output voltage also changes by a corresponding amount so that it is negative with respect to the zero level and recovers with a time constant towards the zero level. Fig. 1.33 shows the pulse shapes. The equation relating the output and input voltages is

$$v_2 = v_1 \frac{sCR}{1 + sCR},$$

for which a solution for a step input 0 to V_1 is,

$$v_2 = V_1 \exp\left(-t/RC\right).$$

Thus the exact form of the output waveform depends on the value of the time constant RC.

28

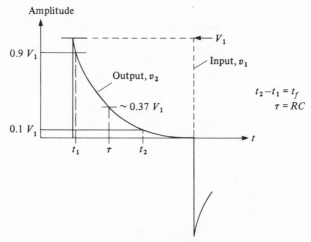

Fig. 1.33. Time constants.

Some of the features of this pulse waveform are worth noting. After a time $t = RC$ the output v_2 is 36.8 per cent of V_1. The output falls from 90 per cent of its maximum value to 10 per cent of its maximum value in a fall-time of $t_f = 2.2RC$.

1.10.6 Square wave applied to the capacitance–resistance network

The response of the circuit to a square wave is shown in fig. 1.33. The shape again depends on the ratio of the period $2T$ and the time constant RC. If the network is to produce a nearly faithful reproduction of the input then RC must be made as large as possible. On the other hand, if RC is made very much smaller than the period the output is in the form of very narrow spikes of signal which are useful for certain applications. Fig. 1.34 shows a range of wave shapes that are obtained as the time constant is varied from $RC = 0.1T$ to $RC = 10T$.

The circuit is known as a *differentiating* circuit since the differential of a perfect square step is an infinitesimally short spike and an approximation to this condition may be obtained with the circuit. It is also worth noting that the circuit attenuates the low frequency components of a pulse rather more than the high frequency components and is therefore referred to as a high-pass circuit.

An interesting case arises when the input to the circuit is a square wave with rise- and fall-times, t_1 and t_2, as shown in fig. 1.35(a). If one assumes that at the start of the rise and fall the voltage varies approximately

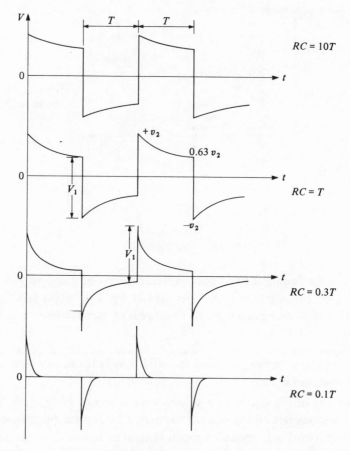

Fig. 1.34. Output waveforms for an ideal square wave applied to a differentiating network. *RC* varies from 10*T* to 0.1*T*.

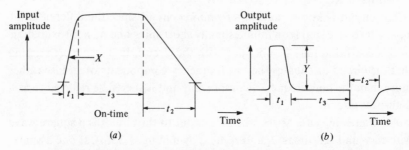

Fig. 1.35. (*a*) Input pulse to differentiating network. (*b*) Output pulse.

linearly with time, then the differentiation of this voltage with respect to time is a constant. A high rate of rise in voltage, X, will give a high output, Y, as shown in fig. 1.35(b). When the input voltage is nearly constant, shown during the on-time t_3, the output is nearly zero and a negative going pulse appears during the fall-time t_2.

1.11 Worked example

A certain voltage amplifier has input resistance $= 100\,\text{k}\Omega$, voltage gain $= 20$, and output resistance $= 5\,\text{k}\Omega$. It is desired to amplify $4\,\text{mV}$ RMS signals coming from a source of internal resistance $5\,\text{k}\Omega$ and to develop at least $20\,\text{mW}$ in headphones of resistance $= 1\,\text{k}\Omega$. Hence find:

(a) How many stages of amplification are needed?

(b) What value of coupling capacitor is needed between the last amplifier stage and the headphones if the output can drop to $10\,\text{mW}$ at $30\,\text{Hz}$?

(c) What length of screened cable can be used between the source and the first amplifier stage if the capacity between conductors in the cable is $60\,\text{pF/metre}$ and if the output can be allowed to drop to $10\,\text{mW}$ at $16\,\text{kHz}$?

Fig. 1.36 shows n amplifier stages between source v_i and load R_4.

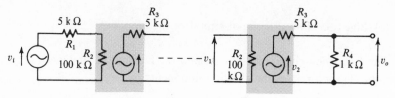

Fig. 1.36 Circuit for example 1.11(a).

(a) Considering first the output stage, power at the output $= v_o^2/R_4$ and must be $20\,\text{mW}$. Hence:

$$v_o^2 = 20 \times 10^{-3} \times 10^3 = 20 \quad \text{so} \quad v_o = 4\cdot47 \text{ volts RMS.}$$

In the output stage, we have from (1.1):

$$\frac{v_o}{v_2} = \frac{R_4}{R_3 + R_4} \quad \text{and} \quad \frac{v_2}{v_1} = 20.$$

So for n stages, (1.2) gives:

$$\frac{v_o}{v_i} = 20^n \left(\frac{100\,\text{k}\Omega}{100\,\text{k}\Omega + 5\,\text{k}\Omega}\right) \left(\frac{100\,\text{k}\Omega}{100\,\text{k}\Omega + 5\,\text{k}\Omega}\right)^{n-1} \left(\frac{1\,\text{k}\Omega}{1\text{k}\Omega + 5\,\text{k}\Omega}\right) \geqslant \frac{4.47}{4.0 \times 10^{-3}}.$$

By inspection, three stages are required and with $n = 3$,

$$v_i = \frac{4.47 \times 1.05^3 \times 6}{20^3} = 3.9 \text{ mV gives an output of } 20\,\text{mW.}$$

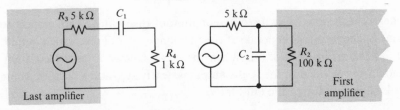

Fig. 1.37. Circuit for example 1.11 (b)   Fig 1.38. Circuit for example 1.11 (c).

(b) Fig. 1.37 shows the output circuit with the coupling capacitor in series with the load. The power of 10 mW expected at 30 Hz is half that expected for higher frequencies and so 30 Hz is the half power frequency. Equation (1.20) then relates the circuit components by

$$1/\omega C_1 = R_3 + R_4.$$

Hence: $$C_1 = \frac{1}{\omega(R_3 + R_4)} = \frac{1}{2\pi \times 30 \times 6000} = 0.88 \times 10^{-6} \text{ F.}$$

So a 1 μF capacitor would be suitable.

(c) Fig. 1.38 shows the input circuit of the amplifier.

The frequency given, 16 kHz, is again the half power frequency. Equation (1.21) then relates the circuit components by:

$$\frac{1}{\omega C_2} = \frac{R_1 R_2}{R_1 + R_2} = \frac{5 \text{ k}\Omega \cdot 100 \text{ k}\Omega}{105 \text{ k}\Omega} = 4.76 \text{ k}\Omega.$$

$$C_2 = \frac{1}{2\pi \times 16 \times 10^3 \times 4.76 \times 10^3} = 2.10 \times 10^{-9} \text{ F} \quad \text{or} \quad 2100 \text{ pF.}$$

The cable capacity is 60 pF/m so 35 metres of cable would give exactly the above capacity and should be the maximum length between the source and the first amplifier stage.

1.12 Summary

Fig. 1.12 shows the required ratio of load to source resistances to get good current, voltage or power coupling from one circuit to another. The load is not always adjusted to get the best power matching. If it is known that a device has a linear relation, say, between its output and input currents, then it is desirable that the source be loaded so that current is efficiently coupled into the device. This is precisely the case with most bipolar transistor devices and circuits containing them require different coupling criteria from circuits containing field-effect transistors if the best linearity is to be obtained.

Fig. 1.22 shows how the reactive components most commonly present in amplifiers and in their input and output coupling circuits affect the gain of the circuit at extremes of frequency. The drop in gain at low frequency can be avoided by a little extra circuit complexity but that at high frequencies cannot be avoided. In the graph of gain magnitude against frequency, the relationship approximates to one broken into lines of different slopes. The frequencies at which the breaks occur can be easily calculated and form the basis for determining the extent of the band where the gain is reasonably constant. Such calculations are good enough for most practical design work when we remember the considerable tolerances given by the makers of transistors and capacitors on the values of their components.

1.13 Problems

1. A voltage amplifier is shown in fig. 1.39. In testing the amplifier, the following results were obtained for a constant input, $v_1 = 0.1$ volts RMS from a source of negligible internal resistance:

   (*a*) Output, $v_4 = 8.0$ volts RMS when $R_1 = 0\ \Omega$ and $R_4 = \infty$.
   (*b*) Output, $v_4 = 4.0$ volts RMS when $R_1 = 0\ \Omega$ and $R_4 = 1\ \text{k}\Omega$.
   (*c*) Output, $v_4 = 2.0$ volts RMS when $R_1 = 1\ \text{M}\Omega$ and $R_4 = 1\ \text{k}\Omega$.

   Hence find the input resistance R_2, voltage gain A, and output resistance R_3 of the amplifier.

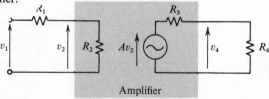

Fig. 1.39. Circuit for problem 1.

2. The output circuit of an amplifier can be represented by a current generator I in parallel with a resistor R_5 as shown in fig. 1.40.

   Derive an expression for the output current, i_4. What relation of R_5 to R_4 is required for the output current to be within 1 per cent of I?

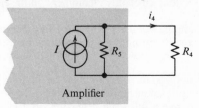

Fig. 1.40. Circuit for problem 2.

33

If this circuit is of the same amplifier as that in question 1, what are R_5 and I in terms of R_3 and Av_2?

3. The circuit of an amplifier is shown in fig. 1.41. Calculate the magnitude of the gain, $|v_4/v_2|$ at 1 Hz, 1 kHz and 1 MHz. What is the amplifier bandwidth between half-power points?

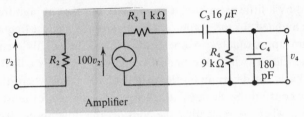

Fig. 1.41. Circuit for problem 3.

4. For the circuit shown in fig. 1.40 derive an expression for the power developed in R_4. If I and R_5 are fixed, what value of R_4 makes this power a maximum?

In a certain power amplifier, $R_5 = 10$ ohms and $I = Gv_2$, where $G = 100$ S ($= A/V$) and v_2 is the amplifier input voltage. What value of load resistor R_4 is desirable, and what input voltage will cause 2.5 watts output to be developed? If the load has to be joined to the amplifier through a coupling capacitor what value is needed for this component if the output voltage of the circuit is only to drop to 70 per cent of normal as the frequency is lowered to 30 Hz?

5. An amplifier is made from three stages whose performance at 1 kHz is as follows:

	Input resistance	Gain	Output resistance
Stage A	5 Ω	30	15 Ω
B	100 Ω	100	500 Ω
C	10 MΩ	4 mS	10 kΩ

A and B are bipolar transistor stages and their current gains are given; C is a field-effect transistor stage and the gain given is the transconductance of the stage.

In what order should the stages be connected to make a voltage amplifier capable of driving a 15 Ω load efficiently? Comment on the efficiency of coupling between stages.

For an input to the amplifier of 0.1 V RMS from a generator of 600 Ω internal resistance, what approximate output power would be developed in the load?

(Sheffield University: Second year)

6. A designer has available standard amplifiers of input resistance 100 ohms, current gain 50 and output admittance 10^{-3} S. He wants to make a circuit to detect 1 μA DC and to give a full scale deflection on a 1 mA DC meter whose resistance is 200 Ω.

How many stages of amplification does he need? What shunt would he put across the input to give exactly the desired sensitivity?

By adding a capacitor to the input, he limits the gain of the circuit for frequencies above 10 Hz. Show how this capacitor should be connected and determine a suitable value for it.

(Cambridge University: First year)

2

The p–n junction and the field-effect transistor

2.1 Introduction

The p–n junction is used a great deal in semiconductor circuitry, both on its own and as an integral part of many amplifying semiconductor devices. As a circuit element its most frequent use is as a diode or rectifying component. It is also used in the form of a Zener diode to act as a voltage reference and its non-linear characteristic is utilised in detector circuits. In integrated circuits, p–n junctions are found as diodes, resistors and capacitors! Both field-effect transistors and bipolar transistors contain one or more p–n junctions. We therefore begin this chapter with a discussion of the p–n junction although our basic intention in this book is to restrict ourselves to amplifying devices and their circuits. The reader who is not familiar with terms such as p-type and n-type material, doping, impurities, etc., which are used in discussing semiconductor devices can find a brief account in the books listed in appendix A.

The operation of the field-effect transistor, its characteristics and its use as a voltage amplifier are described. Representation of the transistor by an equivalent circuit of passive components and generators leads to a method of analysis for gain, input and output impedance. The metal-oxide semiconductor transistor is also described in this chapter. Finally, a numerical example is solved to illustrate circuit design and analytic techniques.

The sections on field-effect transistors may be read before or after the next chapter on bipolar transistors depending on the wishes or needs of the reader. The explanations in the two chapters have been kept reasonably independent of each other to allow the student this choice.

2.2 The p–n junction

A p–n junction diode is formed when an intimate contact or junction is made between a p-type and an n-type semiconductor. In actual devices the 'junction' is a result of doping the material in such a manner that there is

36

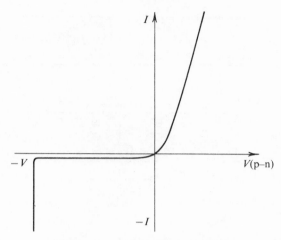

Fig. 2.1. Current flow in a p–n junction.

a transition from p-type to n-type material in a very short distance, typically less than 1 μm. The main property of the device is that current flowing across the junction increases rapidly with applied voltage when the *p-type* side is made *positive* with respect to the n-type side. If the polarity is reversed and the voltage increased, a very much smaller and almost constant current flows until, at a critical voltage, the device breaks down and a large reverse current flows. This last part of the characteristic is known as the avalanche or Zener breakdown. Fig. 2.1 illustrates the current flow as a function of the polarity of the voltage on the p-type side.

The operation of the device is explained by considering the movement of holes and electrons in a region in the immediate vicinity of the junction. Fig. 2.2(*a*) is a schematic representation of the p- and n-type materials as they would exist in isolation. Only the impurity atoms are shown and the lattice of Ge or Si atoms is not included in the interests of clarity. For the p-type side, each impurity atom is shown as an immobile nucleus with a negative charge together with a *positively charged hole*† which is taken to be free to move. In the n-type side donor atoms are shown as immobile nuclei, each with a positive charge together with a *negatively charged* free electron.

When a voltage is applied to an n-type semiconductor the greater part of the current flow is caused by the movement of free electrons which are therefore called the *majority carriers*. However, there are always some holes, albeit comparatively few, present in n-type material. They are called

† The concept of holes is explained in books listed in appendix A for readers unfamiliar with the terminology used to describe the action of semiconductor devices.

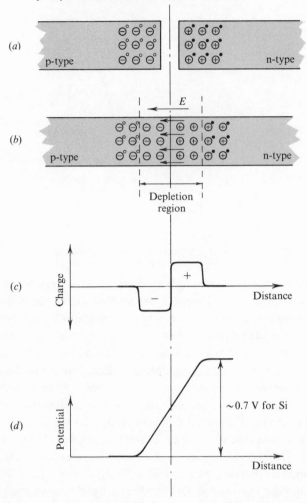

Fig. 2.2. The p–n junction. (*a*) Impurities in semiconductors: ● electrons, ○ holes, ⊖ bound acceptor atoms, ⊕ bound donor atoms. (*b*) The depletion region at a p–n junction. (*c*) The space-charge at a p–n junction. (d) potential barrier at the junction.

the *minority carriers* and they contribute a small fraction to the total current. The situation in p-type material is reversed in that holes form the majority carriers and electrons the minority carriers.

The materials, as a whole, are electrically neutral. In other words, the number of positive charges is balanced by the number of negative charges and there is no *net* charge.

Fig. 2.2(*b*) shows the situation when a p–n junction is formed. The holes in the p-type region, the *majority* carriers, are in much greater concentration than in the adjacent n-type region, hence there is a *diffusion* of holes into the n-type region where they are assimilated into the lattice by recombination with electrons. (Other texts describe the mechanism of hole and electron diffusion and recombination.) Immobile impurity atoms with a net negative charge are left in the p-type side of the junction. In a similar manner, the diffusion of *majority* carriers, i.e. electrons from the n-type region, creates a net positive charge in this side of the junction. Fig. 2.2(c) shows, schematically, the charge distribution in the region around the junction, known as the *space-charge* region. The region is also referred to as the *depletion layer* because of the relative scarcity in it of current carrying charges. It must again be realised, that despite the formation of the space-charge region the overall charge neutrality of the device is maintained.

Once the space-charge region has been formed it restricts further diffusion of holes and electrons. Very few electrons from the n-type side are able to overcome the negative charge on the p-type side of the junction and very few holes are able to overcome the positive charge on the n-type side of the junction. However, small numbers do cross, and if this were the only charge movement in the p–n junction, a small current from p-region to n-region would flow. But there are always some *electrons* on the *p-type* side and some *holes* on the *n-type* side and for these electrons and holes the space-charge region certainly does not act as a barrier. Instead they are helped across from one side to the other once they enter the space-charge region. The current arising from these charge carriers is in the direction n-region to p-region and exactly balances the current from the charges which surmount the barrier. A state of dynamic equilibrium always exists and there is no *net* current flow.

The creation of the space-charge regions means that an electric field is produced across the junction as shown by the arrows in fig. 2.2(b). The existence of the field implies that a potential barrier exists at the junction as illustrated in fig. 2.2(d). It is this potential difference which gives a measure of the barrier which holes and electrons have to overcome to cross the junction. The magnitude of the potential barrier in a typical silicon p–n junction is in the range 0.5 to 0.9 V and for a germanium junction 0.2 to 0.4 V.

2.3 Forward and reverse biased p–n junctions

Forward bias may be applied to a p–n junction by connecting the *positive* terminal of a battery to the *p-type* side and the negative terminal to the n-type side. The immediate effect of this is to reduce the height of the potential barrier. The positive terminal connected to the p-type side repels holes into the negatively charged impurity atoms near the junction, neutralises some of these, and therefore reduces the magnitude of the space-charge. Similarly the negative terminal of the battery causes more electrons to move into the space-charge region on the n-type side and neutralise some of the positively charged impurity atoms near the junction. The magnitude of the potential barrier is reduced and eventually it is altogether eliminated as the applied voltage is increased. The effect of this connection is to inject holes from the p-type side into the n-type side and to inject electrons in the opposite direction and therefore cause a considerable current flow.

Reverse bias is applied when the p-type side is made negative with respect to the n-type side. This increases the width of the space-charge region and the magnitude of the potential barrier. Holes are extracted from the p-type region and electrons from the n-type region, with the result that more immobile ionised atoms are exposed around the junction. The height of the potential barrier is increased to such an extent that the diffusion component of current is reduced to negligible proportions. The current through the diode is only the component from thermally created (minority) charge carriers and its magnitude is very small.

The relation between the applied voltage and the diode current is as follows:

$$I = I_s \left(\exp \frac{eV}{kT} - 1 \right), \tag{2.1}$$

where I is the diode current, V is the applied voltage, e is the electronic charge, k is Boltzmann's constant and T is the absolute temperature.

At room temperature the value of $e/kT \approx 40$ V^{-1}. Therefore, an applied voltage of only -0.1 volt in the reverse direction means that effectively a current of $-I_s$ flows. This is the current caused by thermally generated electrons and holes. Again if a forward bias is applied to the junction, a voltage of only $+0.1$ volt is enough to cause a current of approximately $50I_s$ to flow because $\exp 4 \approx 54$.

Fig. 2.3 shows the forward characteristics of a BY126 silicon diode for two values of junction temperature. Some of the main details from the manufacturer's data are also included.

2.4   Depletion layer width and junction capacitance

The width of the depletion layer at a p–n junction depends on the doping levels in the semiconductor and on the voltage applied to the p–n junction.

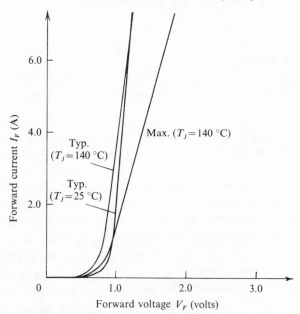

Fig. 2.3. The forward characteristic of a BY 126 silicon diode. Peak reverse volts $= 650\,V$, reverse current $= 0.08\,\mu A$ at $V_R = 450\,V$, $T = 25\,°C$.

Many semiconductors utilise these properties in their operation and detailed analytic discussion of the topics may be found in books on the physics of semiconductor devices. In this section we present a qualitative discussion of the subject.

If the two sides of the p–n junction are doped uniformly and the doping levels are equal, then the extent of the depletion layer will be the same on either side of the junction. This may be deduced from the fact that the crystal remains electrically neutral.

Very often the doping is such that one side of a p–n junction is much more heavily doped than the other. The result is that the depletion layer extends much further into the lightly doped side than into the heavily doped side. Since charge neutrality must be maintained, equal amounts of charge must be exposed on both sides of the junction. A greater volume of the lightly doped side would be required than of the heavily doped side for any given number of ionised impurities. This creates an asymmetrical depletion region extending into the lightly doped side. It is in fact this type of junction which is used in many semiconductor devices.

The effect of a voltage applied to the p–n junction has already been mentioned in §2.3. A reverse bias causes an increase in the width of the depletion region and forward bias causes a decrease. This effect is used in

41

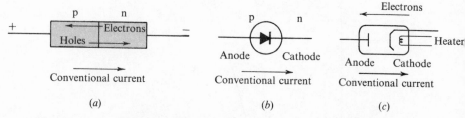

Fig. 2.4. (*a*) Movement of holes and electrons and conventional current direction in a p–n junction. (*b*) The diode symbol. (*c*) The vacuum diode analogy.

devices such as the field-effect transistor to control the current. In addition the junction may be regarded as two conducting regions with a region of high resistivity separating them, which is essentially the structure of a capacitor. The magnitude of the capacitance depends on the reverse voltage applied to the junction because this varies the depletion layer width. Therefore not only can we form a capacitor by means of a p–n junction, but also the value of this capacitor is a function of the applied voltage and variable capacitance diodes are useful devices for circuit tuning and other applications.

The capacitance at a p–n junction can have a detrimental effect in some devices, e.g. in bipolar transistors, where it limits the response at high frequencies. On the other hand it does provide a convenient way of making capacitors for integrated circuits as will be seen in a later chapter.

Fig. 2.4(*a*) shows the direction of current flow in a p–n junction and fig. 2.4(*b*) shows the conventional symbol used to denote a p–n junction diode. The direction of current flow is indicated by the arrowhead if forward bias is applied.

Fig. 2.4(*c*) shows the origin of the anode and cathode terminology which is based on the thermionic diode. It also helps one to remember the direction of conventional current flow.

The characteristics of the diode are markedly non-linear. In the reverse direction the effective resistance of a silicon diode is usually greater than $10^8\ \Omega$. The current remains virtually constant with applied reverse voltage until breakdown. In the forward direction the resistance is a function of the point on the characteristic at which it is measured.

Since

$$I = I_s \left(\exp \frac{eV}{kT} - 1 \right)$$

$$\frac{\mathrm{d}I}{\mathrm{d}V} = \frac{e}{kT} I_s \exp \left(\frac{eV}{kT} \right) \approx \frac{e}{kT} I,$$

Depletion layer and junction capacity

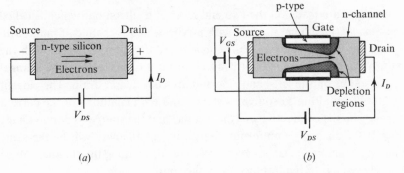

(a)   (b)

Fig. 2.5. (a) Current through a silicon channel. (b) The junction gate field-effect transistor (schematic representation).

and therefore
$$\frac{\mathrm{d}V}{\mathrm{d}I} = r = \frac{kT}{eI} \approx \frac{1}{40I}\,\Omega$$

since $e/kT \approx 40$ at room temperature.

r is called the incremental or dynamic resistance of the diode. Its value varies from a few kilohms when the diode current is a few μA, to a few ohms when an appreciable current is being passed by the diode.

We have dealt mainly with the silicon diode in this account. The characteristics of germanium diodes are similar except that the reverse current is much higher than for a silicon diode, typically a few μA rather than a few nA. In the forward direction a lower forward bias than for a silicon diode causes the device to conduct.

2.5 The field-effect transistor (unipolar transistor)

The central component of any amplifier circuit is an active device. One such device is the field-effect transistor, so called because an electric field is used to vary the flow of current in it. Another is the MOST or metal oxide semiconductor transistor described in §2.17. The devices are also known as unipolar transistors because, in the main, their mechanism of operation depends on the movement of charge carriers of one polarity only, either holes or electrons. This is in contrast with the bipolar transistor (described in chapter 3) which depends on the movement of both holes and electrons.

2.6 Principle of operation

The operation of an f.e.t. can be visualised by considering the somewhat schematic device geometry illustrated in fig. 2.5. Although this geometry is not used in practical devices it provides an adequate model from which we may discuss the behaviour of the device.

Contacts are made on the two end faces of a silicon bar which is called the *channel* (n-channel if made with n-type Si and p-channel if made with p-type Si). The contacts are called *source* and *drain* respectively. With the battery connections as in fig. 2.5(*a*), electrons are injected into the channel at the source on the left-hand side and are collected at drain. The current I_D flows from drain to source as shown and the magnitude of the current depends on the conductivity of the channel and the source to drain voltage V_{DS}. If the channel is uniformly doped, its conductivity will be the same at all points along its length and the voltage drop along the channel will be linearly related to the distance from the source.

The action of the f.e.t. depends on providing a means of varying the conductivity of the channel and hence the current I_D. Two p-type regions are diffused into opposite transverse sides of the silicon bar to form two p–n junctions which are known as *gates* – see fig. 2.5(*b*). Let us assume the gates to be reverse biased so that depletion layers are formed around the p–n junctions. If the n-channel is lightly doped compared with the p-type gates, the depleted regions will extend well into the channel and decrease its *effective* width as far as the conduction of current is concerned. It has been pointed out in §2.4 that the conductivity of the depletion layer is very low because there are very few carriers in it. The decrease in the effective channel width implies an increase in its resistance and a decrease in the channel current.

Let us now consider the action in more detail and derive the characteristics of a f.e.t. Consider the gate voltages to be zero initially, i.e. the gates are connected directly to the source. Drain current flow causes a voltage drop along the length of the channel. So the gate to channel voltage is a small reverse bias towards the source end of the channel and a much larger reverse bias at the drain end. Hence a wedge-shaped depletion layer forms as shown in the figure. As V_{DS} is increased from zero, current I_D increases linearly initially. This part of the characteristic is marked A in fig. 2.6. Further increase of V_{DS} however causes the channel to narrow and its resistance to increase. The effect of the increased resistance begins to dominate over the increase in I_D which would have been expected, and the characteristic begins to form a 'knee' marked B in fig. 2.6. A state of equilibrium is reached when there is no further increase in I_D. The wedge-shaped depletion layers extend across the channel and it is almost closed at the drain end. This point is called pinch-off and is marked C in fig. 2.6. For higher voltages than the pinch-off voltage, V_{PO}, nearly constant current flows and the part of the characteristic is marked D in fig. 2.6. Any increase of V_{DS} and extension of the depletion layer is counter-balanced by the

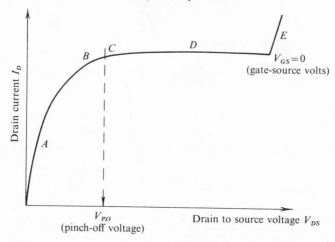

Fig. 2.6. The I_D-V_{DS} output characteristic of a f.e.t. with $V_{GS} = 0$.

high field in the narrow channel allowing saturation current to flow. At still higher values of V_{DS}, usually well beyond the manufacturer's maximum values, there is a gate to drain breakdown caused by avalanche effects and f.e.t. action ceases. This part of the characteristic marked E shows the current increasing rapidly with V_{DS}.

So far we have considered the gate to source voltage, V_{GS} to be zero. If the gate is made negative with respect to the source, the argument still applies except that pinch-off and saturation occur at a lower value of drain current. A depletion layer is formed and the channel width reduced, even before V_{DS} is applied. The increased resistivity of the channel leads to the change in the slope of the I_D-V_{DS} characteristic in the region before pinch-off. Since the channel is already narrowed pinch-off occurs at lower values of V_{DS} as well as lower drain current. The set of curves obtained with V_{GS} as a parameter is shown in figure 2.7(a). The locus of the pinch-off points is also shown and is known as the pinch-off curve.

The characteristics of the f.e.t. may also be displayed by a plot of I_D against V_{GS} as shown in fig. 2.7(b). This is known as the transfer characteristic and holds for values of V_{DS} greater than the pinch-off voltage. Since the drain current is nearly constant beyond pinch-off a single characteristic is sufficient and is independent of V_{DS}.

The device described in the foregoing section is said to work in the *depletion mode*, in other words a negative gate voltage is used to extend the depletion region and constrict the channel to produce pinch-off. It is possible to envisage a device in which, initially, the channel is partially closed with $V_{GS} = 0$ and the drain current is low. The gates of such a

45

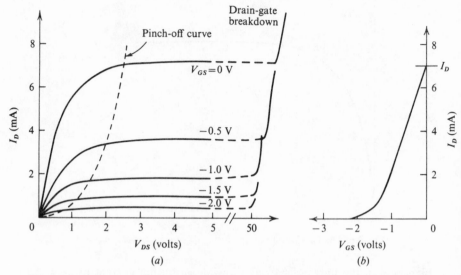

Fig. 2.7. (*a*) The output characteristics of a f.e.t. as a function of V_{GS} and the pinch-off curve. (*b*) The I_D–V_{GS} transfer characteristic of a f.e.t.

device may be biased positive to open up the channel wider, such that pinch-off occurs for greater values of drain current. Such a device works in the *enhancement mode*. Note, however, that the forward biased p–n junction would conduct, and this may overload the signal source which is usually applied to the gate–source terminals. Devices may also be constructed to work over a range which includes both the depletion and the enhancement mode. However it is the depletion mode which is of practical importance, and most junction-type f.e.t.s are designed to work in this mode to take advantage of the very high input resistance of the reverse biased gate.

The description of the f.e.t. was based on a channel made with n-type material but it applies equally to a device made with p-type material, except that holes rather than electrons would form the current carriers. The type of device used also determines the polarities of the voltages to be applied to the several electrodes. The symbols used for f.e.t.s are shown in fig. 2.8. Note that the direction of the arrow on the gate electrode determines whether the device has a p-type or an n-type channel. The n-channel device shows the arrow at the gate pointing *into* the device and the p-channel·device shows the arrow at the gate pointing *out* of the device. The arrows are not meant to indicate a direction of current flow, but follow the conventions employed in the p–n junction symbol where the arrow pointed from the p- to the n-region. For the n-channel device,

46

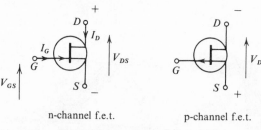

n-channel f.e.t.   p-channel f.e.t.

Fig. 2.8. The symbols for f.e.t.s.

normally, supply voltages are required so that the drain is *positive with respect to source* and the gate is *negative with respect to source*. For the p-channel device the polarities are reversed. Fig. 2.8 shows the conventional symbol for an n-channel f.e.t. with the various currents and voltages marked on it. The figure defines the meaning of these quantities for subsequent sections of this chapter.

2.7 Characteristics of the field-effect transistor

We return to the characteristics of the f.e.t. to discuss the important question of what limits must be placed on the various voltages that may be applied to the device. Fig. 2.6 has already shown an obvious limitation where the device breaks down from avalanche effects. This region of the characteristic is normally to be avoided and V_{DS} is kept well below the avalanche value. However, before the voltage reaches this value, we may well encounter a limiting condition based on the maximum power which may be dissipated in the device. We have two sources of power dissipation, one at the gate and one between drain and source. Thus total power dissipated is:

$$P = I_G V_{GS} + I_D V_{DS}.$$

Now $I_G V_{GS}$ is very small since I_G is extremely small, typically a few nanoamps. Thus we can write that:

$$P \approx V_{DS} I_D.$$

Manufacturers specify the maximum permissible value of power which may be a fraction of a watt for small signal devices. This concept may be conveniently represented on the characteristics of fig. 2.9. Since $I_D = P/V_{DS}$ or for a given value of P_{max}, $I_D = \text{const.}/V_{DS}$, we obtain a locus in the form of a hyperbola on the characteristics. Each point on the hyperbola represents the maximum value of $I_D \times V_{DS}$ which must not be exceeded. Another way of visualising the limitation is to say that the region marked beyond, i.e. to the right of the hyperbola, is a forbidden region of operation

47

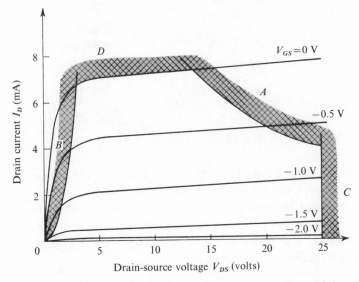

Fig. 2.9. The forbidden regions of operation (shown for 2N3819 output characteristics).

(region A on fig. 2.9). Another region which is avoided in the operation of linear amplifiers is that marked B in fig. 2.9 where the characteristics are so unevenly spaced.

Yet another limit is the maximum voltage which is set by the manufacturer. This voltage must not be exceeded if breakdown of the device is to be avoided, and its maximum value defines a forbidden region C on the characteristics of fig. 2.9. Finally, a fourth forbidden region arises because usually one cannot operate this device in the condition where current may be drawn from the gate, as the input resistance is then low. This constraint implies that for an n-channel device V_{GS} must be zero or negative. Thus we have the fourth forbidden region D, shown on the characteristics of fig. 2.9.

2.8 Field-effect transistor amplifier

Before we consider how we may use the f.e.t. as an amplifier of fluctuating voltages, or signals, we have to consider the steady state or DC conditions which must be provided for the device. We wish to know how to choose a supply voltage and a load, bearing in mind that we have to avoid the forbidden regions of the characteristics. If we do not fulfil this requirement the device may overheat and fail, or it may give a highly distorted output signal.

Field-effect transistor amplifier

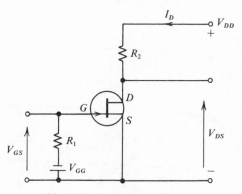

Fig. 2.10. The f.e.t. amplifier.

The simplest possible arrangement is shown in fig. 2.10. A power supply, V_{DD}, is connected to the drain, D through a resistor R_2, which we will call, for reasons which will become apparent later, the load resistor. Another power supply V_{GG} is connected between source and gate via a resistor R_1. This is known as the *bias* supply. Under these conditions of steady supply voltages V_{GG} and V_{DD}, we have the following relations:
At the input or gate terminal,

$$V_{GG} = I_G R_1 + V_{GS}. \tag{2.2}$$

At the output terminals,

$$V_{DD} = I_D R_2 + V_{DS}. \tag{2.3}$$

For (2.2) we have $I_G \approx 0$ since the device draws very little current. Therefore

$$V_{GG} \approx V_{GS}.$$

For (2.3), I_D and V_{DS} are both variables and the equation represents a relation which is plotted on the characteristics as shown in fig. 2.11. Differentiating (2.3) with respect to I_D gives: $dV_{DS}/dI_D = -R_2$, a linear relation. It may also be seen that the intercepts of this line on the two axes are the points A and B given by:

$$V_{DS} = 0 \quad \text{at} \quad I_D = \frac{V_{DD}}{R_2} \quad \text{point } A,$$

$$I_D = 0 \quad \text{at} \quad V_{DS} = V_{DD} \quad \text{point } B.$$

This line on the characteristics is known as the *load line* and is a locus of the only values of I_D and V_{DS} that can exist together once V_{DD} and R_2 have been chosen. The choice of V_{DD} and R_2 must be undertaken with care to ensure that the load line lies on the allowed side of the power dissipation curve shown in fig. 2.9. We can obtain any combination of I_D and V_{DS}

49

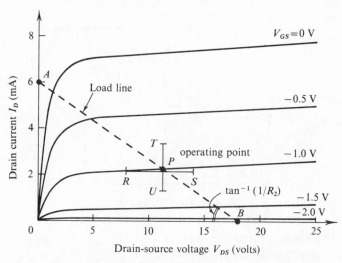

Fig. 2.11. The load line and the operating point (shown for 2N3819 output characteristics).

given by the load line by selecting a third parameter, the voltage V_{GS}. Let us consider the extreme cases first. If V_{GS} is made zero or positive for an n-channel device we would be in the region where appreciable I_G would exist and usually this is avoided. On the other hand if V_{GS} is made excessively negative we would be in the region where little or no drain current would flow. The amplifier would tend to distort the signal to be amplified and therefore this region of operation is avoided. In the characteristics shown in fig. 2.11, there would be very little drain current for V_{GS} more negative than -2 V.

It is advisable to choose V_{GS} to avoid the forbidden regions of operation by a good margin of safety. Once V_{GS} is chosen the DC or steady operating conditions of the f.e.t. are fixed at a point on the load line such as the one marked P on fig. 2.11. This point is known as the 'quiescent' *operating point*. As an example, note that as shown in fig. 2.11, the operating point P and the load line fulfil the following conditions:

Supply voltage $V_{DD} = 18$ V.
The load resistor $R_2 = 3$ kΩ.
Drain to source voltage $V_{DS} = 11.2$ V.
Bias voltage $V_{GG} = -1$ V.
Drain current $I_D = 2.2$ mA.
Power dissipation in the device $V_{DS} \times I_D = 25$ mW.

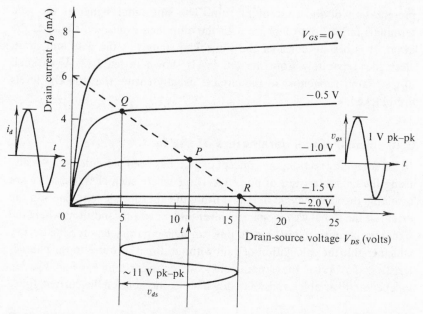

Fig. 2.12. Graphical presentation of gain (shown for 2N3819 output characteristics).

2.9 Amplification of signals

Having set the DC conditions on the f.e.t., in other words having established the operating point P, consider how a fluctuating voltage applied between the gate and source terminals will be amplified by the f.e.t. circuit of fig. 2.10. Let the signal voltage be a sine wave, with a peak to peak amplitude of 1 volt as shown in fig. 2.12. The existence of the signal means that the instantaneous value of the gate voltage will vary from its original steady value of -1 volt to the two extremes -0.5 volt to -1.5 volts, shown by points Q and R. This disturbance should not be so large as to take the device close to the non-linear region of its characteristics.

The drain current would fluctuate as shown in fig. 2.12, and under these conditions the voltage across the f.e.t. would also fluctuate as predicted by the load line. If this voltage fluctuation at the drain is compared with the fluctuation in V_{GS}, we realise that an amplified version of the input has been obtained at the drain terminal. From Q to R the voltage V_{DS} changes from $+5$ V to $+16$ V, the total change in V_{DS} being 11 V for the 1 volt input. Note that in this explanation the working points have all been taken along the load line so that the conditions linking V_{DS} and I_D are always met. One other point must be made in this graphical

51

presentation of the concept of gain. The sinusoidal signal at the gate terminals is 180° out of phase with the amplified sinusoidal signal at the drain. It is best to consider the sinusoidal signals at the gate and drain electrodes separately from the DC levels shown in fig. 2.12. As we shall discuss later, capacitors in the circuits usually ensure that the DC levels are blocked.

2.10 Small signal parameters of the f.e.t.

In the foregoing section, the concept of gain has been presented graphically using the characteristics of fig. 2.12. Thus, where such characteristics are available the gain may be obtained using the appropriate load line and the extent of the signal variations. However under certain conditions where the signal variations are kept small, one can define parameters for the device, which enable the calculation of gain without direct recourse to the characteristics. To derive these parameters we consider that V_{GS} and V_{DS} are independent variables and write down an expression for the current I_D:

$$\Delta I_D = \left| \frac{\partial I_D}{\partial V_{GS}} \right|_{V_{DS}} \Delta V_{GS} + \left| \frac{\partial I_D}{\partial V_{DS}} \right|_{V_{GS}} \Delta V_{DS}. \qquad (2.4)$$

This equation may be interpreted that the change in I_D is produced by small changes, ΔV_{GS} and ΔV_{DS}. These small changes are taken around the operating point P, which we assume is well within the allowed region of operation and on a fairly evenly spaced part of the characteristics. For these conditions we can give a meaning to the parameters:

$$\left| \frac{\partial I_D}{\partial V_{GS}} \right|_{V_{DS}} \quad \text{and} \quad \left| \frac{\partial I_D}{\partial V_{DS}} \right|_{V_{GS}}.$$

The latter, $|\partial I_D / \partial V_{DS}|_{V_{GS}}$ is the slope of the *characteristic* at the point P. It is called the drain conductance and given the symbol g_d. Its inverse, $r_d = 1/g_d$, is more often quoted and referred to as the drain resistance. $|\partial I_D / \partial V_{GS}|_{V_{DS}}$ is the change in I_D corresponding to a change in V_{GS}, along a vertical line drawn through the point P, i.e. with V_{DS} kept constant. It is the mutual or trans-conductance and given the symbol g_m. These two parameters are, strictly speaking, defined only for the operating point P. They are *not constant* and will vary from point to point on the characteristics. For example, the slope of the characteristics is not constant and indeed on approaching the pinch-off point would vary considerably. Also the spacing of the characteristics decreases for equal increments of V_{GS} as the gate voltage becomes more negative. Despite these constraints g_m

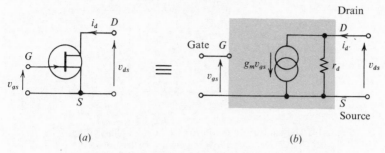

Fig. 2.13. (*a*) The f.e.t. and (*b*) its small-signal equivalent circuit.

and r_d are widely quoted and are used to predict the behaviour of the circuit under conditions where the signals are kept small and restricted to the vicinity of the operating point P.

For example, from the characteristics of fig. 2.11 we may obtain the values of r_d and g_m in the region of the operating point P. We take an excursion from R to S along the $V_{GS} = -1$ V characteristic. Then V_{DS} rises by 6 V and I_D by 0.2 mA, whence $r_d = 30$ kΩ. Keeping V_{DS} constant we have a change I_{UT} of 2 mA in drain current for a change of 0.5 V in gate to source voltage. Therefore $g_m = 4$ mA/volt or 4 mS (milli-siemens).

Using these parameters, r_d and g_m, we may rewrite (2.4) as:

$$i_d = g_m v_{gs} + \frac{v_{ds}}{r_d}. \tag{2.5}$$

Here the lower case symbols v_{ds}, i_d and v_{gs} have been substituted for the small changes ΔV_{DS}, ΔI_D and ΔV_{GS}. This equation describes the behaviour of the f.e.t. over a restricted range. It also enables us to represent the f.e.t. by an equivalent circuit using voltage and current sources and resistors. The f.e.t. and its equivalent circuit are shown in figs. 2.13(*a*) and (*b*). At the input side on the left, assume that a voltage v_{gs} is applied. At the output side on the right, assume we have a current generator $g_m v_{gs}$ with a resistor r_d in parallel. The output voltage is v_{ds}. This is a synthesis of the equation in terms of electrical components. As a check we write down an expression for the current i_d flowing into the drain node. It must equal the sum of the currents flowing out of the node, through the current generator and the drain resistance, r_d. Since the voltage across r_d is v_{ds} the latter current is v_{ds}/r_d, therefore we have:

$$i_d = g_m v_{gs} + \frac{v_{ds}}{r_d}$$

53

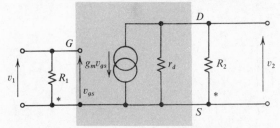

Fig. 2.14. The equivalent circuit for the f.e.t. amplifier of fig. 2.10.
* Denotes a low impedance battery.

which is (2.5) once again. Once the equivalent circuit is known we have a method of representing the f.e.t. in any amplifier circuit, so that we may conveniently analyse the circuit. The analysis usually requires a calculation of the gain, the input and output impedances and the frequency response.

For the simplest of amplifier circuits such as the one shown in fig. 2.10, we obtained a gain of approximately 11 by the graphical technique of §2.9. In fig. 2.14 we show the same amplifier with the f.e.t. replaced by its equivalent circuit which is done as follows. Looking in turn at each terminal of the equivalent circuit, we add the components connected to that terminal in fig. 2.10. Thus to the gate we have to connect the input signal and the resistor R_1 as these were the only paths connected in fig. 2.10. Connections to each of the other terminals of the equivalent circuit are shown similarly. All batteries or power supplies are treated as short circuits for the signal, because even though they do have some resistance associated with them, it is very small compared with the other resistances in the circuit and is regarded as insignificant for our purposes. The part of fig. 2.14 which represents the equivalent circuit of the f.e.t. is shown in the grey tinted area.

2.11 The common-source amplifier (AC analysis)

The first interesting application of the equivalent circuit is to obtain a relation between v_1 and v_2, the input and output voltages respectively for the circuit of fig. 2.10 which is known as the common-source amplifier. The term common-source is used because the source terminal is shared by both the output and input sides of the amplifier. For the input side we have $v_1 = v_{gs}$ (fig. 2.14). This step is a seemingly trivial one in this case but in some applications, the source of input voltage, v_1 may have a very high internal resistance, and therefore v_{gs} will be less than v_1. One has then to use (1.1) to find the value of v_{gs} for a source voltage, v_1. At the output

54

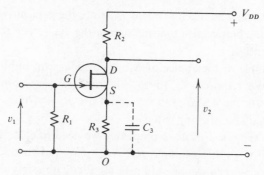

Fig. 2.15. Biasing of a f.e.t. amplifier.

side, the current $g_m v_{gs}$ flows through the parallel combination of r_d and R_2. Hence the voltage v_2 is given by:

$$v_2 = -g_m \frac{R_2 r_d}{R_2 + r_d} v_1 \quad \text{since} \quad v_1 = v_{gs}.$$

The gain
$$\frac{v_2}{v_1} = -g_m \frac{R_2 r_d}{R_2 + r_d}. \tag{2.6}$$

The magnitude of the amplifier gain is therefore $|g_m R_2 r_d/(R_2 + r_d)|$. The minus sign in (2.6) shows that v_2 is $180°$ out of phase with v_1. This corresponds with the result which was pointed out earlier in the graphical presentation, that v_{gs} and v_{ds} are out of phase by $180°$. Substitution of the numbers for g_m and r_d (obtained in §2.10) and for R_2, gives a gain, $|A| \approx 11$. Any difference between this value and that from the graphical analysis may be accounted for by the assumption that g_m and r_d are constants.

2.12 Simple bias circuit for the f.e.t.

The use of a battery between gate and source to provide the bias V_{GS} has to be avoided in practical circuits. A simple means is to use the circuit of fig. 2.15. Here a resistance R_3 is inserted between the source and an arbitrary zero reference level, O, which is often the ground connection. The resistance R_1 ensures that there is a connection between gate and O even when the input signal generator is disconnected. The gate to source voltage $V_{GS} = V_{GO} - V_{SO}$, where V_{GO} and V_{SO} are the gate to reference and source to reference respectively.

A typical value for the resistance R_1 is ~ 1 MΩ and the gate current is very small, say of the order of 10^{-9} A. $V_{GO} = I_G R_1 \approx 10^{-3}$ V, and can be neglected in comparison with the usual values of V_{GS} which may be of the order of 1 V. Thus the expression for the bias voltage is:

$$V_{GS} = -V_{SO} = -I_{SO} R_{SO} = -I_D R_3.$$

55

The current I_D may be taken to equal I_{SO} since the gate current is assumed to be negligible. Notice also that the sign of the bias is correct, as we want V_{GS} to be negative.

As an example of the choice of R_3 we consider our amplifier circuit of §2.8 again. We required V_{GS} to be -1 V and I_D was 2.2 mA. Therefore $R_3 = 1/2.2 \times 10^{-3} = 454\ \Omega$, and we would use $R_3 = 470\ \Omega$ as the nearest commonly available resistor value.

To specify the bias resistor completely we need to calculate the power dissipated in it. In this example $I^2R = 22$ mW and therefore a resistor with a $\frac{1}{10}$ watt power rating would give an adequate safety margin and be suitable for our amplifier.

The resistor, R_3, therefore, ensures that the desired bias of -1 V exists between gate and source. Unfortunately a consequence of introducing R_3 into the circuit is to decrease the gain obtainable from the amplifier. Analysis using a circuit similar to fig. 2.14 but with the resistance R_3 included between source and ground gives a gain less than that given by (2.6). However, where the signal to be amplified is not a steady or very low frequency signal, there is a simple and effective means of overcoming this disadvantage. A capacitor, known as a decoupling capacitor, is placed in parallel with R_3 as shown in fig. 2.15. The capacitor, if it is correctly specified, effectively bypasses the resistance R_3 for the AC signal and yet maintains a steady bias between gate and source.

The value of the capacitor is determined by using the criterion that its impedance must be much less than the other impedances between source and ground. At first sight it appears that it would be sufficient to make $1/\omega C_3 \ll R_3$ where ω is the lowest frequency of interest. In fact analysis shows that the impedance looking into the source of the transistor is approximately $1/g_m$ and the correct value for C_3 is to make its impedance less than ($1/g_m$ in parallel with R_3) at the lowest frequency of interest.

2.13 Input and output impedance

It has been pointed out that the input impedance of a field-effect transistor is very high because it is effectively the resistance of a reverse biased p–n junction. It is usually in the region of 10^9 ohms and the equivalent circuits, such as the one in fig. 2.14, which do not show any impedance other than R_1 between gate and source are to a great extent justified. At high frequencies, however, this statement is no longer true and capacity between gate and source C_{gs}, and more particularly capacity between drain and gate have a significant effect. This is more fully discussed in §2.15 on high frequency

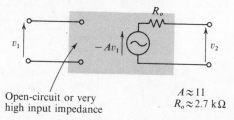

Open-circuit or very
high input impedance

$A \approx 11$
$R_o \approx 2.7\,\mathrm{k}\Omega$

Fig. 2.16. The Thévenin equivalent circuit of a f.e.t. amplifier.

behaviour of f.e.t.s. The high input impedance makes the f.e.t. a suitable device for the amplification of voltage signals rather than current signals. Analysis similar to that used in §1.2 is applicable to f.e.t. amplifiers. Note also that the dependent generator in the equivalent circuit of fig. 2.13 is specified in terms of the input voltage, v_{gs}.

For the common-source amplifier of fig. 2.10 the output circuit is a current generator $g_m v_{gs}$ with parallel resistors r_d and R_2; so the *output* resistance is clearly given by:

$$R_o = \frac{R_2 r_d}{R_2 + r_d}. \tag{2.7}$$

This expression is valid over a considerable frequency range but at both extremes of very low and very high frequencies it must be modified.

At midband frequencies the f.e.t. amplifier may be regarded as a voltage amplifier such as shown in fig. 2.16. For our amplifier circuit of fig. 2.10 with the 3 kΩ load resistor we had $|A| \approx 11$ and now we have the output resistance as 3 kΩ in parallel with 30 kΩ or

$$R_o \approx 2.7\ \mathrm{k}\Omega.$$

2.14 Coupling of f.e.t. amplifiers, and amplification of low frequency signals

The basic principles common to the coupling of amplifiers have been described in chapter 1. Now we present the circuit of an f.e.t. amplifier with such coupling components at its output terminals as are necessary to connect it to subsequent stages of amplification. In the circuit shown in fig. 2.17(a), R_4 is the gate to source resistor of the following transistor which ties the gate to ground potential. C_4 blocks the steady potential at the drain $V_{DS}(+11\ \mathrm{V})$ from upsetting the gate to source bias of the following transistor. The impedance of C_4 should be low enough so that the signal output of the first stage, v_3, is not appreciably less than the signal output at the drain, v_2. The equivalent circuit of fig. 2.17(a) is shown in fig. 2.17(b). The f.e.t. is replaced by the part of the circuit within the grey

57

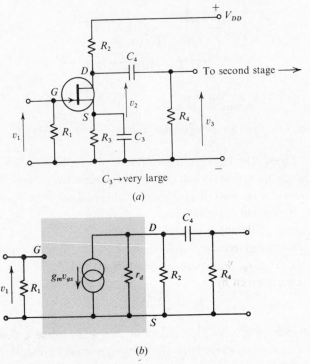

Fig. 2.17. (a) Coupling a f.e.t. amplifier to a second stage through a R–C network. (b) Equivalent circuit for fig. 2.17(a).

tinted area. This circuit is similar to that discussed in §1.8 on coupling circuits at low frequencies, and the gain as a function of frequency is:

$$A = \frac{-g_m R'}{1 + R'/R_4 + 1/j\omega R_4 C_4},$$

where

$$R' = \frac{r_d R_2}{r_d + R_2}.$$

R'/R_4 is usually $\ll 1$.

Therefore,

$$A \approx \frac{-g_m R'}{1 + 1/j\omega R_4 C_4} = \frac{A_0}{1 + 1/j\omega R_4 C_4}, \tag{2.8}$$

where A_0 is the gain without the coupling capacitor C_4. As has been pointed out earlier, in chapter 1, the gain drops as the frequency is decreased and there is a phase shift between input and output. The frequency at which A drops to $0.707 A_0$, the output voltage v_3 also drops to $1/\sqrt{2}$ of the midband value, and this is known as the half power point. At this point the frequency is given by

$$\omega = 1/R_4 C_4 \quad \text{or} \quad f = 1/2\pi R_4 C_4.$$

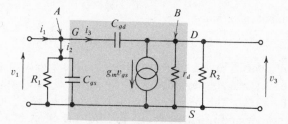

Fig. 2.18. Junction capacities, gate to drain and gate to source.

Note also that the phase angle between input and output would be 45° at this particular frequency. For f.e.t. amplifiers, the value of R_4 would usually be in the region of 1 MΩ and if good audio frequency response is required a coupling capacitor of 0.01 μF would give a half power frequency of about 16 Hz. Note the close similarity of (2.8) and (1.10) of chapter 1, and the general conclusions obtained in that chapter apply to the f.e.t. amplifier.

In this discussion, the impedance of R_3 in parallel with C_3 has been assumed to be very small. In a simple analysis, the effects of C_3 and C_4 can be examined independently and separate calculations can be made to determine suitable values for them. One of the capacitors, whichever is cheaper or smaller, can be made say, ten times larger than calculated, so that the remaining capacitor effectively determines the low frequency response.

2.15 Amplification of high frequency signals

At high frequencies we must take into account the capacity between gate and source and the capacity between gate and drain. These capacities arise from the p–n junction at the gate and it will be shown that the latter is the more significant of the two. The equivalent circuit must now be modified as shown in fig. 2.18. We ignore the coupling capacity at the output side because at high frequencies this does not have an appreciable effect. This circuit is best analysed by considering the voltage at the two nodes A and B. We have

$$v_A = v_1, \quad i_2 = v_1(\mathrm{j}\omega C_{gs}) + v_1/R_1,$$

$$v_B = Av_1 \quad \text{and} \quad i_3 = (v_1 - Av_1)\,(\mathrm{j}\omega C_{gd})$$

$$= v_1(1 - A)\,(\mathrm{j}\omega C_{gd}).$$

Total input current

$$i_1 = v_1[\mathrm{j}\omega C_{gs} + (1 - A)\,(\mathrm{j}\omega C_{gd}) + 1/R_1]. \tag{2.9}$$

59

Hence the input *admittance*,

$$Y_1 = i_1/v_1 = j\omega C_{gs} + (1-A)j\omega C_{gd} + 1/R_1. \tag{2.10}$$

This is equivalent to saying that the input circuit capacity of the f.e.t. at high frequencies may be regarded as a capacitor C_{gs} in parallel with a capacitor $(1-A)C_{gd}$. The magnitude of the two capacitors, C_{gs} and C_{gd}, are of the same order and usually about 1 pF. But the effect of C_{gd} is much more pronounced because of the multiplication by the factor $(1-A)$. This reduction of the input impedance of the f.e.t. at high frequencies is the most serious aspect of the existence of C_{gd} and C_{gs}; it tends to diminish the advantages of the high resistive component of input impedance which makes the f.e.t. an attractive device for low and medium frequency operation.

A reduction of gain also arises as a result of stray capacity, and the overall frequency response of the amplifier is typically as described in chapter 1. There is a fall off at both high and low frequency ends of the frequency spectrum.

In addition to the capacity effects at the junctions the frequency response is affected by the transit time of the carriers, which is taken into account by makers who quote a reduced or complex value for g_m at high frequencies.

†2.16 The source follower

An important amplifier circuit using a field-effect transistor is known as the source follower (also known as the common-drain amplifier). The circuit is shown in fig. 2.19(a) for an n-channel f.e.t. The drain terminal is connected directly to the positive rail of the power supply while the source is connected via a resistor R_1 to the negative rail which is also the common rail between input and output voltages v_1 and v_2. The resistors R_2 and R_3 are used to provide the correct bias voltage at the gate. The special features of this circuit are that it exhibits a very high input impedance and a low output impedance compared with the common-source amplifier described earlier. However, the voltage gain is less than unity.

The equivalent circuit for the source follower is shown in fig. 2.19(b) where it is particularly important to note the position of the resistor R_1 between source and common rail. The voltage v_{gs} which must be used in specifying the current generator of the equivalent circuit is not equal to v_1; instead:

$$v_{gs} = v_1 - v_2, \tag{2.11}$$

and at the node S we have:

$$g_m(v_1 - v_2) = \frac{v_2}{r_d} + \frac{v_2}{R_1} + i. \tag{2.12}$$

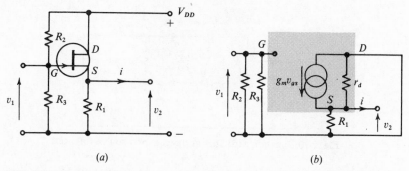

Fig. 2.19. (*a*) The source follower. (*b*) Equivalent circuit of the source follower.

Hence
$$v_2 = \frac{g_m v_1}{1/r_d + 1/R_1 + g_m} - \frac{i}{1/r_d + 1/R_1 + g_m}. \qquad (2.13)$$

It follows therefore that the no-load voltage gain, i.e. when $i = 0$ is

$$\frac{v_2}{v_1} = \frac{g_m}{1/r_d + 1/R_1 + g_m}. \qquad (2.14)$$

It should be noted that the input and output voltages are in phase for the source follower which may be contrasted with the 180° phase shift found in the case of the common-source amplifier.

Usually the drain resistance of the f.e.t., r_d, is much larger than R_1. In that case

$$\frac{v_2}{v_1} \approx \frac{g_m R_1}{1 + g_m R_1}. \qquad (2.15)$$

For typical values it is also often true that $g_m R_1 \gg 1$. Therefore v_2/v_1 is just less than 1.

Note that (2.13) is of the form $v_2 = $ no-load gain $\times v_1 -$ output resistance $\times i$. Therefore, from (2.13) we have:

Output resistance $R_o = \dfrac{1}{1/r_d + 1/R_1 + g_m}.$

The significant properties of the source follower are best illustrated by means of a numerical example, using a set of typical values. For example, with $g_m = 5$ mS, $R_1 = 10$ kΩ and $r_d = 100$ kΩ:

The gain . $\dfrac{v_2}{v_1} = \dfrac{5}{5 \cdot 11} = 0.978$ (from (2.14)).

The output resistance, $R_o = \dfrac{10^3}{5 \cdot 11} = 196$ Ω,

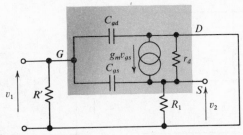

Fig. 2.20. Junction capacities at the gate of a source follower.

which is less than the value obtained in the case of the common-source amplifier, §2.13.

The input resistance has been assumed infinite for the f.e.t. in the equivalent circuit of fig. 2.19(*b*) since the gate does not draw any current. The bias resistance chain of R_2 and R_3 at the gate reduces the input resistance to $R_2 R_3/(R_2+R_3)$. In addition to the input resistance there is also a capacity C_{gs} between gate and source and another capacity C_{gd} between gate and drain as for the common-source amplifier described in §2.15. The effective capacity can be calculated from the equivalent circuit of fig. 2.20. The input admittance:

$$Y_{in} = \frac{i_1}{v_1} = j\omega C_{gd} + j\omega(1-A)\,C_{gs} + \frac{1}{R'},$$

where
$$R' = \frac{R_2 R_3}{R_2+R_3}.$$

Since A is very nearly unity the effective input capacity is $C_{in} \approx C_{gd}$ which is much less than in the case of the common-source amplifier. Hence the high frequency performance of the source follower is better than that of the common source amplifier.

2.17 The metal oxide semiconductor transistor (MOST)

The metal oxide semiconductor transistor (MOST) is the most widely used device in digital integrated circuits and it is particularly useful in making high density circuits such as semiconductor memories. It is also found as a discrete device and it is sometimes designed into the input-end circuitry of analogue circuits such as op-amps. Its circuit operation is very similar to that of the junction field-effect transistor discussed earlier so that it is sometimes called a MOSFET. It has a very high resistance between the gate and the channel so that it is also known as the insulated-gate f.e.t. (IGFET).

There are four basic types of MOSTS: enhancement-mode n-channel and

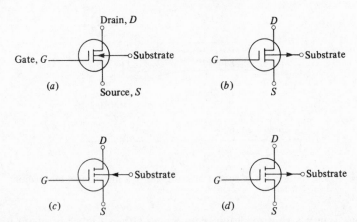

Fig. 2.21. Symbols used for four different types of metal-oxide semiconductor transistors (MOSTS): (*a*) n-channel, depletion-mode, (*b*) p-channel, depletion-mode, (*c*) n-channel; enhancement-mode, (*d*) p-channel, enhancement-mode.

p-channel and depletion-mode n-channel and p-channel devices. The enhancement-mode devices are normally OFF and a voltage has to be applied to turn them ON while the depletion-mode devices normally conduct, or are ON and a voltage has to be applied either to turn them OFF or to vary the current flowing in the device. The symbols for the four types are shown in fig. 2.21. Note that the arrowhead direction on the substrate indicates whether the device is n-channel or p-channel. The gap between the gate and the channel, shown in the symbols, indicates a high resistance between the gate and channel.

The cross-section of a MOST is shown in fig. 2.22. Two regions of n+, or heavily doped silicon, the *source* and *drain* are made in p-type material. The gap between them is known as the *channel*; of length L in fig. 2.22. It

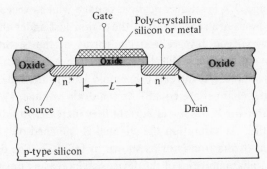

Fig. 2.22. Cross-section of NMOS transistor. An n-channel device is shown. For a p-channel device, PMOS, the p+ regions are made instead of n+ and the substrate is n-type silicon.

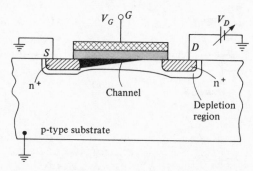

Fig. 2.23. Cross-section of NMOS transistor showing the channel pinched off.

is covered with a thin oxide layer on which either a layer of metal or of conducting poly-crystalline silicon is deposited. This part of the device is known as the *gate*. The figure shows the structure of an enhancement-mode device but in the depletion-mode device the structure is very nearly the same. The channel under the gate oxide is made to be of the same semi-conductor type as the source and drain except that it is relatively lightly doped n-type. For p-channel enhancement-mode devices the substrate is n-type silicon and the source and drain are doped to be p-type while for depletion-mode p-channel devices the channel is also lightly doped p-type.

The action of the device is very similar to that of the field-effect transistor and again only one polarity of charge carrier, electrons for n-channel devices or holes for p-channel devices, is involved in the conduction process. The action of the device is illustrated in fig. 2.23. Initially with the gate voltage, $V_G = 0$, the source voltage, $V_S = 0$ and with the drain voltage V_D at a low positive voltage, no current flows between the source and drain since there are reversed biased p–n junctions at their boundaries. When the gate voltage, V_G, is set at a positive voltage which exceeds a threshold voltage, V_T, holes are repelled from the region just under the gate oxide, and a narrow n-type layer effectively forms at the Si–SiO$_2$ interface, which allows conduction of electrons from source to drain or a current flow drain to source. (At the exact threshold voltage the hole and electron densities in the channel are equal.) As the drain voltage, V_D, is increased from a low value for $V_G > V_T$ the current flow increases rapidly at first and finally saturates. At saturation the channel is 'pinched-off' at a position near the drain of the transistor as shown in fig. 2.23. The drain current, I_D, versus V_{DS} characteristics of the device for increasing positive values of V_G are shown in fig. 2.24(a) for an n-channel, enhancement-mode transistor. The characteristics of a depletion-mode device are shown in fig. 2.24(b),

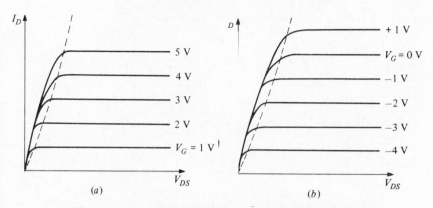

Fig. 2.24. MOST characteristics: (*a*) n-channel, enhancement-mode, (*b*) n-channel, depletion-mode.

in which it may be seen that, in this case, there is current flow when $V_G = 0$. In a depletion-mode n-channel device in which the region just under the gate oxide is lightly doped n-type there is a connection between the two n$^+$ regions even with $V_G = 0$. When V_G is made positive the saturated drain current increases while for negative values of V_G the drain current decreases until for large negative values the drain current ceases to flow altogether. In this condition the voltage at the gate repels sufficient electrons from the channel to make it effectively p-type. The input/output characteristics for enhancement-mode and depletion-mode devices are of the form shown in fig. 2.25(*a*) and (*b*). Note that the threshold voltage has to be exceeded before the enhancement-mode device conducts while for the depletion-mode device there is a current flow for both negative and positive values of V_{GS}. The slope of the characteristic gives the forward conductance of the device; namely the current change in the output for a voltage change in the input. This is the main effect that has to be modelled in devising the small signal equivalent circuit for the device. In MOSTs the oxide layer at the gate, which is a good insulator, isolates the gate electrode from the channel. As a consequence there is a high input resistance at the gate for both positive and negative polarities of V_G, whereas with f.e.t.s the gate resistance is low when the voltage applied at the gate is such that the input p–n junction is forward-biased. It should be noted that the fourth electrode shown in the device symbol is the substrate which is generally connected to earth potential and this is assumed here. In some applications substrate bias is used. but these considerations are beyond the scope of this book.

The small-signal equivalent circuit for the MOST is the same as that for the f.e.t. shown in fig. 2.13(*b*) and the analysis of analogue MOST circuits

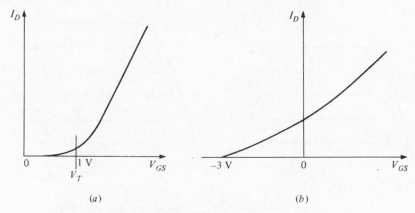

Fig. 2.25. MOST characteristics: (*a*) n-channel, enchancement-mode, (*b*) n-channel, depletion-mode.

follows the procedures outlined earlier in this chapter. Digital applications of this device are more important than analogue or linear applications and these are considered in chapter 7.

2.17.1 Complementary MOS (CMOS) transistors

In complementary MOS circuitry both n-channel and p-channel transistors are constructed in the same substrate in close proximity to each other. Fig. 2.26 shows the cross-section of adjacent devices. The n-channel device is constructed in a well of p-type material created in an n-type substrate while the p-channel device is constructed in the substrate itself. Guard rings may be placed around the device to prevent lateral leakage currents. Alternatively n-wells can be created to house p-channel devices.

The transistors are generally connected in series as shown in fig. 2.27. The gates are joined together to form the common input terminal while the

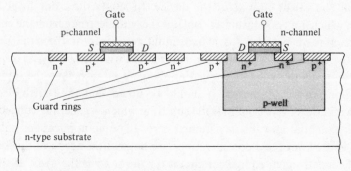

Fig. 2.26. Cross-section of CMOS transistor structure.

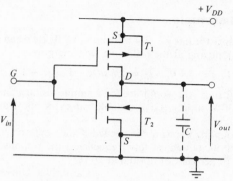

Fig. 2.27. Circuit of CMOS transistor switch.

output is taken from the common drain connection. In DC conditions the application of a voltage V_{in} = HIGH turns T_2 ON and T_1 OFF so the output voltage V_{out} is LOW. Since T_1 is OFF there is no continuous path between V_{DD} and earth so no direct current flows. When V_{in} is LOW T_2 is OFF and T_1 is ON and the output voltage is HIGH. Again there is no direct connection between V_{DD} and earth. Thus the output switches between HIGH and LOW but there is no direct current flow path in either of these steady states. There is only a very small leakage current and the power taken from the supply in steady conditions is negligibly small.

During the switching transient, however, power is drained from the supply. The output capacitance, C (broken line in fig. 2.27) must be charged up by the supply when the transistor T_2 is OFF and discharged to earth when the transistor goes to LOW. Assuming that the output voltage swing is very nearly from earth to V_{DD} we can calculate the power dissipation. The energy stored in the capacitance is $\frac{1}{2}CV_{DD}^2$, and the capacitor is discharged rapidly to earth once in each cycle. The power dissipation is $CV_{DD}^2 \times f$ watts in each transistor when the frequency is f. Thus the power dissipation in CMOS circuits increases with operating speed. The steady power dissipation when the circuit is not switching is as low as a few nanowatts. The circuit is insensitive to spurious signals or noise and operates satisfactorily over a wide voltage range. It is widely used in digital circuitry. The disadvantages are that CMOS circuitry occupies a larger area of silicon than NMOS circuitry, it operates at lower frequency and is generally more expensive. CMOS circuitry is widely used in digital watches, calculators, microcomputers and other portable electronic equipment.

2.18 Worked example

A field-effect transistor has $g_m = 4$ mS and $r_d = 10$ kΩ when connected in the common-source mode and at the following working point:

$$V_{DS} = +8 \text{ V}, \quad I_D = 2.5 \text{ mA}, \quad V_{GS} = -2 \text{ V}.$$

Draw the circuit that you would use for a simple voltage amplifier using this transistor and a 30 V supply. Give suitable values in ohms and watts for the resistors in your circuit.

What output coupling capacitor is needed if the next amplifier stage has an input resistance of 500 Ω and, at low frequencies, the response can be allowed to drop to about 70 per cent of normal at 20 Hz?

(Cambridge University: First year)

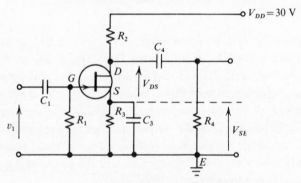

Fig. 2.28. Circuit for worked example.

A suitable circuit is shown in fig. 2.28.

(i) *To determine R_1 and the bias resistor R_3.* R_1 ties gate to earth potential and is typically 1 MΩ. The current through R_1 is small, say 2 nA, and the power dissipated is very low. Therefore a low power, 0.1 watt, resistor would be suitable.

$$R_1 = 1 \text{ MΩ}, 0.1 \text{ watt}.$$

$$R_3 = \text{source-base resistor} = V_{SE}/I_D$$

$$= 2 \text{ V}/2.5 \times 10^{-3}$$

$$= 800 \text{ Ω}.$$

The nearest commonly available value is 820 Ω.

$$\text{Power dissipation } I^2 R_3 = (2.5)^2 \times 10^{-6} \times 800$$

$$= 5 \text{ mW}.$$

Therefore use 0.1 watt resistor again.

$$R_3 = 820 \text{ Ω}, 0.1 \text{ watt}.$$

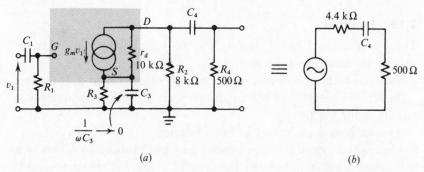

Fig. 2.29. (*a*) Equivalent circuit for worked example. (*b*) Thévenin equivalent for fig. 2.29 (*a*).

(ii) *To determine the load resistor R_2.*

$$R_2 = \frac{V_{DD} - V_{DS} - V_{SE}}{I_D} = \frac{30 - 8 - 2}{2.5 \times 10^{-3}}$$

$$= 8 \text{ k}\Omega.$$

Commonly available value $R_2 = 8.2 \text{ k}\Omega$.

$$\text{Power dissipation } I^2R = (2.5)^2 \times 10^{-6} \times 8 \times 10^{-3}$$

$$= 50 \text{ mW.}$$

Therefore use 0.1 watt resistor.

$$R_2 = 8.2 \text{ k}\Omega, 0.1 \text{ watt.}$$

Note that commonly available values of resistors have been given and the power dissipation has been checked to be within the rating of miniature 0.1 watt resistors.

(iii) *To determine the value of the coupling capacitor.* Draw an equivalent circuit for the output side of the amplifier (fig. 2.29(*a*)). Assume that C_3 is very large so its impedance is negligible at 20 Hz. The analysis follows the steps outlined in §1.8. The circuit may be simplified to that shown in fig. 2.29(*b*). If the output at 20 Hz is to drop to 70 per cent of the value at midband frequency then from (1.12)

$$\frac{1}{\omega C} = 4.4 \text{ k}\Omega + 500 \ \Omega.$$

Putting $\omega = 2\pi \times 20$ we have:

$$C = 1.6 \ \mu\text{F.}$$

2.19 Summary

The p–n junction forms a vital part of most amplifying semiconductor devices. Its main properties are:

(a) it conducts current more easily in one direction than in the other;

(b) depletion regions of low conductivity are formed in the immediate vicinity of the junction;

(c) a capacity is associated with the junction.

The conductivity of a silicon channel may be controlled by means of a p–n junction and this leads to the concept of the field-effect transistor which is an amplifying device whose characteristics are shown in fig. 2.6. To use the device as an amplifier one has first to establish suitable operating conditions by choosing a supply voltage and the load and bias resistors. Once the operating conditions are adequately specified the amplification of relatively small electrical signals is achieved as demonstrated graphically in fig. 2.11.

Two small signal parameters are obtained from the characteristics and are called the mutual conductance, g_m and the drain resistance, r_d. They are used to derive an equivalent circuit of the device which represents it for purposes of analysis. The gain, input impedance and output impedance are calculated using the equivalent circuit.

The coupling capacitor used to link the f.e.t. amplifier to a second stage or to a load limits the gain at low frequencies. An inadequate bypass capacitor across the biasing resistor in the source can also have the same effect. At high frequencies the input impedance is reduced by the junction capacitances, particularly by that between gate and drain.

The source follower is another useful amplifier circuit. Although its voltage gain is less than unity its input impedance is very large and its output impedance is less than that of the common-source amplifier.

The alternative to the junction gate f.e.t. is the MOST which has the property of a very high input impedance which is independent of the polarity of the voltage applied to the gate.

2.20 Problems

1. A field-effect transistor has a small signal equivalent circuit with input resistance = 1000 MΩ, forward transfer conductance = 4 mS and output conductance = 100 μS when at the operating point, $V_{DS} = +4$ V, $I_D = 2$ mA, $V_{GS} = -2$ V. ·

Draw the circuit that you would use for a single stage voltage amplifier. Describe the use and specify the value of as many components as possible if a 30 V supply was available.

What voltage gain would you expect when the output was unloaded? Give reasons which might account for not getting this gain exactly.

(Sheffield University: First year)

2. A field-effect transistor is used as a voltage amplifier and with a load resistor of 40 kΩ a gain of 40 is obtained. If the load resistance is halved, the voltage gain drops to 30. Calculate the output resistance and the mutual conductance of the transistor.

Briefly compare the advantages and limitations of the field-effect transistor with the bipolar transistor and with the pentode.

(Sheffield University: First year)

3. A field-effect transistor is used as a voltage amplifier with a load resistor of 45 kΩ. When the load resistor is halved the output resistance is reduced to 91 per cent of its original value. Find the voltage gain for both values of the load. The mutual conductance is 6×10^{-3} A/V.

(Sheffield University: First year)

4. What is meant by the terms: midband gain, upper and lower cut-off frequencies, and bandwidth when applied to a wideband, capacitively coupled amplifier?

In fig. 2.30 the capacitance C_{ds}, measured between source and drain, is 6 pF and the stray wiring capacitance, C_s, is equivalent to a 4 pF capacitor in parallel with the 25 kΩ load.

Determine the midband gain of the amplifier, expressed in dB, and the frequencies at which the gain is 3 dB below this midband value.

Sketch the frequency response of the amplifier.

(Birmingham University: First year)

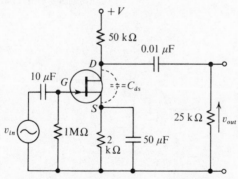

Fig. 2.30. Circuit for problem 4. $g_m = 2000\,\mu S$, $g_d = 20\,\mu S$.

5. Describe the principle of operation of a symmetrical n-channel field-effect transistor. Sketch a typical set of I_D–V_{DS} characteristics, and account for their shape in the various regions.

An audio amplifier stage incorporates a drain resistance of 3.3 kΩ in series with an f.e.t., and a source resistance of 470 ohms, all across a 24 volts DC supply. The characteristics of the f.e.t. for $V_{DS} > 3$ volts may be approximated by the following constant current lines:

V_{GS} (volts)	0	-0.5	-1.0	-1.5	-2.0	-2.5
I_D (mA)	7.00	5.00	3.30	1.86	0.78	0

The gate is connected to the negative supply rail through a large-valued resistor, and the source resistor is shunted by a large-valued capacitor. Signal coupling to the gate and from the drain is through adequate capacitors. Find the quiescent DC voltages of the drain and source terminals relative to the negative supply line, and estimate the small signal stage gain, in decibels.

(Bristol University: First Year)

6. Fig. **2.31** shows the circuit, using a field-effect transistor, of one stage of a voltage amplifier.

Obtain an expression for the voltage gain of the stage when supplying a resistive load of value R_3 ohms connected between A and B, assuming that stray capacitance is equivalent to a capacitor of value 100 pF connected between drain and source. The impedance of the source bias circuit may be assumed to be negligible over the designed frequency range.

If $R_1 = 1\,\text{M}\Omega$, $R_2 = 10\,\text{k}\Omega$, $R_3 = 1\,\text{k}\Omega$ and $C = 1\,\mu\text{F}$, determine the bandwidth between the -3 dB points on the voltage–frequency output characteristic when the sinusoidal input p.d. is of constant magnitude and variable frequency.

Three stages having the same circuit configuration as shown, but having different values of components, are connected in cascade. When so connected the stages have the following individual half power band limits:

Stage 1   100 Hz to 250 kHz,
Stage 2   75 Hz to 350 kHz,
Stage 3   50 Hz to 450 kHz.

Obtain an expression from which the high frequency half power limit for the complete amplifier may be obtained.

(Cambridge University: Second year)

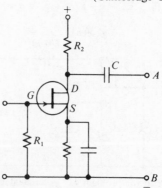

Fig. 2.31. Circuit for problem 6. $g_m = 4\,\text{mS}$, $r_d = 10\,\text{k}\Omega$.

7. Give a qualititative explanation of the operation of a field-effect transistor below pinch-off. Using the characteristics of a field-effect transistor given in

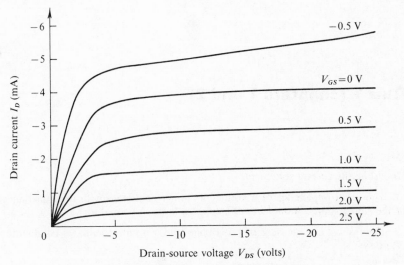

Fig. 2.32. Characteristics for problem 7.

fig. 2.32, design a self-biased, common-source, low frequency amplifier working from a supply voltage of 25 V and having a voltage gain of 10. The amplifier should show less than 10 per cent reduction in gain at a frequency of 50 Hz.

What considerations may cause a limit to be imposed on the magnitude of the alternating voltage swing applied to the gate?

(London University: Second year)

8. An amplifier with a 'common' input and output terminal, and a gain $(-G)$, where G is a positive number, has an admittance Y between the remaining output terminal and input terminal. Determine expressions for the equivalent shunt input and output admittance, stating any assumptions made. Explain the significance of this result.

A multistage amplifier has identical stages, each employing an f.e.t. for which $g = 6$ mS, $r_D = 100$ kΩ, $C_{GS} = 1$ pF, $C_{DG} = 2$ pF, $C_{DS} = 1$ pF. The stage drain resistance R_D is 10 kΩ, and the gate resistance R_G is 300 kΩ. Calculate the stage gain, in dB, and the stage upper cut-off frequency.

(Bristol University: First year)

Quiz 1 (chapters 1 and 2)

Underline the correct statements.

1. The input impedance of a voltage amplifier compared with the impedance of the signal source should be (*a*) small, (*b*) large, (*c*) equal, (*d*) infinite.

2. The decibel is a measure of (*a*) voltage gain, (*b*) power gain, (*c*) bandwidth, (*d*) noise.

3. 'All amplifiers are designed to obtain the maximum power gain' is (*a*) true, (*b*) false.

4. The gain of a *R–C* coupled amplifier (*a*) falls at high frequencies, (*b*) falls at low frequencies, (*c*) remains constant at all frequencies.

5. At the 'half-power point' the voltage gain of an amplifier falls to (*a*) $1/\sqrt{2}$ of the maximum value, (*b*) $\frac{1}{2}$ the maximum value, (*c*) $\frac{1}{4}$ the maximum value.

6. The conventional direction of current is in the direction of flow of (*a*) electrons, (*b*) holes.

7. The p–n junction shown in fig. 1 is (*a*) forward biased, (*b*) unbiased, (*c*) reverse biased.

Fig. 1. Fig. 2.

8. The symbol shown in fig. 2 is (*a*) a p-channel f.e.t., (*b*) a diode, (*c*) an n-channel f.e.t., (*d*) a MOST.

9. Mutual conductance or g_m is defined as

$$(a)\ \left|\frac{\partial I_D}{\partial V_{GS}}\right|_{V_{DS}}, \quad (b)\ \left|\frac{\partial V_{DS}}{\partial I_D}\right|_{V_{GS}}, \quad (c)\ \left|\frac{\partial V_{DS}}{\partial V_{GS}}\right|_{I_D}.$$

10. The high frequency response of a f.e.t. depends mainly upon the value of (*a*) its gate-drain capacitance, (*b*) its gate-source capacitance, (*c*) its drain resistance, (*d*) its g_m.

11. The bias from gate to source in the circuit of fig. 3 is (*a*) 1.0 V, (*b*) 0.5 V, (*c*) −0.5 V, (*d*) −1.0 V.

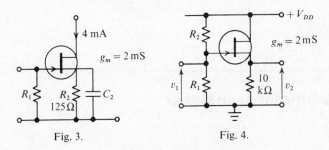

Fig. 3.   Fig. 4.

12. To minimise the effects of feedback in the circuit of fig. 3 the impedance of the capacitor C_2 should be (a) 6.25 Ω, (b) 125 Ω, (c) 12.5 kΩ, (d) 0 Ω.

13. The circuit of fig. 4 is called (a) a source follower, (b) a common-drain amplifier, (c) a common-source amplifier.

14. The voltage gain of the circuit of fig. 4 is (a) approximately 20 times, (b) much less than unity, (c) exactly unity, (d) just less than unity.

3

The bipolar transistor

3.1 Introduction

Although a semiconductor amplifier using a field-effect phenomenon was postulated by Shockley in 1952, it was not successfully made until 1963. A bipolar transistor was devised and made by Brattain and Bardeen in 1948: it has developed from almost individually made devices which were sealed in glass envelopes like little valves to the mass produced, robust, cheap devices that we know today. Many of the present integrated circuits, described in chapter 4, use bipolar transistors as their active elements whether they be switches or amplifiers. Some integrated circuit designs using the field-effect transistor are also available but their higher cost must be offset by definite requirements for low noise or very high input resistance. A more detailed comparison of bipolar with field effect transistors is made in §3.16.

The bipolar transistor is used in the power amplifiers of our domestic sound equipment, in the largest computers, and in the most complex integrated circuits. This chapter describes first the principle of operation of the bipolar transistor and its typical characteristics. Then §§3.6 on will describe its use in simple amplifier circuits and discuss problems such as its biasing, stability of operating point, likely gain and frequency response. Lastly some more advanced circuits are considered and a numerical example is worked through.

3.2 Principle of operation

Consider the n–p–n sandwich of semiconductor shown in fig. 3.1(a). This contains two back-to-back p–n junctions; see §§2.1 to 2.4 if you are not familiar with p- and n-type materials, junctions, leakage currents, etc. The regions are called collector, base and emitter and are also shown as C, B and E on the symbol for the transistor, fig. 3.1(b). The emitter is always differentiated from the collector in this symbol by having an

76

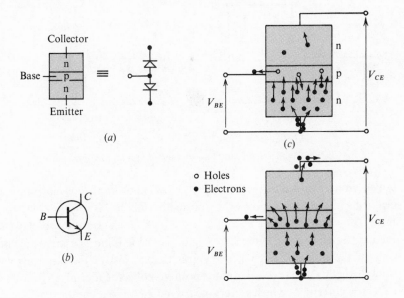

Fig. 3.1. (*a*) The regions of a bipolar transistor, (*b*) its symbol, (*c*) and (*d*) the charge movement in the device.

arrow on it. The reason for the direction of the arrow will be dealt with later.

When the collector region is more positive than either of the other regions, then the collector to base, n–p junction is reverse biased and so only a small leakage current flows. This is the normal operating condition for a transistor of this type, i.e. a *positive* potential on the collector region of an npn transistor applied as V_{CE} in fig. 3.1 (*c*). (If this voltage is reversed, then the collector region changes to a conducting junction and the device operation is quite changed – this will not be considered at present.)

The other applied voltage that is necessary to make the device operate in its normal amplifying mode is one which makes the central region, the base, more positive than the emitter. The base–emitter, p–n, region will then conduct as in an ordinary diode. If the emitter region is heavily doped and the base is lightly doped, the current will consist mostly of electrons, the majority carriers in the n-type material of the emitter, flowing across the base-emitter junction into the base, shown in fig. 3.1 (*c*). There will be a small number of holes, the majority carriers in the base flowing in the opposite direction. This flow of holes gives rise to a component of base current I_{B1}.

Consider now the large flow of electrons that is going from the heavily

77

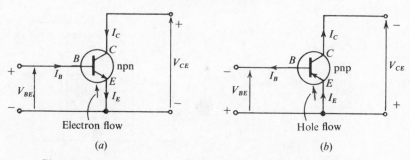

Fig. 3.2. Normal supplies and currents in (*a*) npn and (*b*) pnp devices.

doped emitter and into the base. Fig. 3.1(*d*) shows this concentration of electrons in the base region which normally only has a very few electrons in it. The electrons move away from the high concentration area next to the emitter as shown. If the base region is shaped and is extremely thin, only of the order of a few micrometres thick, most of the electrons will flow through the base to the more positive collector and give rise to the collector current I_C which is very nearly equal to the emitter current I_E. A few electrons from the emitter will flow to the base connection or be neutralised by holes at recombination sites in the base area and these will give rise to a component of base current I_{B2}. The total base current I_B is given by

$$I_B = I_{B1} + I_{B2}. \tag{3.1}$$

The collector current, I_C, is usually many times larger than the base current, I_B. To achieve this,

(*a*) the base doping is much less than in the emitter, so I_{B1} is small, and

(*b*) the base region is very thin, so I_{B2} is small.

Normally, I_C/I_B is of the order of 100 and it may be nearly constant over a wide range of current. It is called the beta, β, the h_{FE} or the static common-emitter current gain of the transistor. The transistor may be regarded as a three-terminal amplifying device in which a small current flowing into the base provides a much larger current flowing into the collector. If a slightly higher voltage is applied to the base, the base current increases sharply because the forward voltage–current relation is exponential, see (2.1). This higher base current is usually linearly related to a higher collector current over a wide range of change.

So far the npn bipolar transistor has been described. A similar mechanism explains the gain of the pnp transistor but as the doping in each region is reversed, the potentials across each junction need to be reversed to get transistor action. Instead of a flow of electrons through the device, there is a flow of holes. Fig. 3.2 shows the symbol of each device and the directions

78

of conventional current flow. Note that the arrow on the emitter in the transistor symbol is in the same direction as the conventional current flow (which is the opposite direction to electron flow.) The npn device is used in most of the descriptions in this chapter because then the circuits are directly comparable with those of the n-channel f.e.t. and the valve. However any circuit can be made to work with a pnp transistor instead of an npn transistor by changing the polarity of the supplies and reversing the connections of the other polarity conscious devices such as diodes and electrolytic capacitors.

3.3 Current relationships

We have explained that ideally for a small base current I_B, a much larger collector current I_C flows. In practical devices, there is also a leakage component of current I_{CEO} as the collector to base is a reverse biased junction. Thus,

$$I_C = \beta I_B + I_{CEO} \tag{3.2}$$

and I_{CEO} is the leakage current being taken from the collector supply.

But the transistor is a three-terminal device and with the direction of the currents shown in fig. 3.2(a) the current out of the device must equal the current in, so

$$I_E = I_B + I_C. \tag{3.3}$$

Equation (3.2) has treated the device as one where the base lead is the input and the collector lead is the output; this is called the common-emitter connection as the emitter terminal is common to both input and output side.

However if we were interested in the amplifier shown in fig. 3.3(a) where we need to know what current to draw out of the emitter to get a given collector current, we can eliminate I_B from (3.2) and (3.3) to give

$$I_E = \frac{I_C}{\beta} - \frac{I_{CEO}}{\beta} + I_C,$$

so

$$I_C = \left(\frac{\beta}{1+\beta}\right) I_E + \frac{I_{CEO}}{1+\beta} = \alpha I_E + I_{CBO}, \tag{3.4}$$

where $\alpha = \beta/(1+\beta)$ and is known as the common-base current gain for the transistor. It can be seen to be always less than unity. This may appear at first sight not to be such a useful amplifier because it has a current gain of less than unity, but its voltage and power gains are considerable and it is used in special circuits such as the cascode amplifier.

The leakage current in the common-base connection I_{CBO} is the ordinary leakage current of a single p–n junction with a reverse bias applied to it,

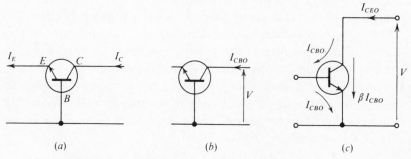

Fig. 3.3. Transistor in common-base circuit and showing leakage currents.

which is shown in fig. 3.3(b). For small p–n junction diodes, it is of the order of a few nA for silicon and a few μA for germanium at room temperature. Re-arranging the leakage current terms of (3.4) give

$$I_{CEO} = I_{CBO}(1 + \beta).\qquad(3.5)$$

It is of interest to be able to visualise how this much higher leakage current arises in the common-emitter connection. Fig. 3.3(c) shows that the leakage current I_{CBO} flowing from the collector to the base must flow on from base to emitter, if the base is unconnected. This base to emitter current gives rise to a current β times more flowing from collector to emitter by normal transistor action. Thus the total leakage current in the common emitter circuit, I_{CEO}, is then the sum of I_{CBO} and βI_{CBO}, as given by (3.5).

The leakage current I_{CEO} is an undesired output in most amplifiers which use the bipolar transistor in the common-emitter connection. Leakage currents are much higher for germanium than for silicon devices, and higher for a power device which has a large junction area. The leakage current rises rapidly with temperature and will increase by more than an order of magnitude for a device temperature change from 25 °C to 85 °C. Thus care must be taken in operating the device if its actual working current is to be precisely achieved.

3.4 Typical common-emitter characteristics

The common-emitter connection has the emitter lead common to the input voltage, V_{BE} and to the output voltage V_{CE}. This is shown in fig. 3.4(a). The input characteristic of the transistor is the relation between current and voltage for the left-hand loop of fig. 3.4(a). It relates the input voltage, V_{BE}, to the input current, I_B, and these are the axes of the typical characteristic sketched in fig. 3.4(b).

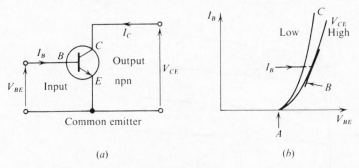

Fig. 3.4. (*a*) Circuit voltages and currents for the common-emitter connection and (*b*) the input characteristic.

The features of importance in fig. 3.4(*b*) are:

A: The characteristic is close to that of a forward biased diode; appreciable current starts to flow at about 0.2 V for germanium devices and about 0.5 V for silicon (see chapter 2).

B: The slope of the characteristic, $\Delta I_B / \Delta V_{BE}$, is the reciprocal of the resistance that the device appears to have if disturbed from some mean base current, I_B. Simple theory gives the resistance in ohms of any forward biased junction as approximately $1/(40 \times$ current in junction) at room temperature. Hence the input resistance can be shown to be about $1/(40I_B)$ ohms. This neglects the resistance of the thin base layer which is justifiable except in high frequency transistors.

C: The characteristic is affected a little by the collector voltage. When the collector voltage is high, a wider depletion layer is made at the collector–base junction. This leaves a thinner conducting base region. In this thinner region, there will be less recombination of holes and electrons; thus to keep the base current constant, a higher base voltage is needed.

The output characteristic of the transistor in the common-emitter connection is a current–voltage relation for the right-hand loop of fig. 3.4(*a*). It relates the output voltage of the circuit, V_{CE}, to the output current, I_C. These are the axes of the typical characteristic sketched in fig. 3.5. Features of importance are:

A: With V_{CE} very low, the collector is not efficient at collecting carriers from the emitter passing through the base region. But after V_{CE} exceeds

81

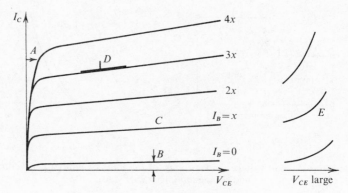

Fig. 3.5. The output characteristic for the bipolar transistor in the common-emitter connection.

a value of a fraction of a volt, known as the 'knee' voltage, the collector current I_C depends largely on the base current I_B and not on the collector voltage.

B: Even with the base current, $I_B = 0$, a small current flows into the collector which is the leakage current, I_{CEO}.

C: For a small base current, $I_B = x$, a considerably larger collector current, I_C flows. I_C/I_B is much greater than unity and is the static current gain as stated earlier. For larger base currents, $I_B = 2x, 3x, 4x$, higher collector currents flow and the characteristics should be evenly spaced for a linear, low-distortion device. Note that the output current is altered nearly linearly by input *current* (unlike the field-effect transistor where output current is related to the input *voltage*).

D: For a given constant base current, $I_B = 3x$ say, note that the characteristic is not horizontal but does rise slightly. Thus as the collector voltage is increased, a little more collector current flows and the device appears to be resistive in the collector circuit. This is a second-order effect (as is the effect of V_{CE} on the input characteristic) and is ignored in approximate calculations of circuit performance.

E: The collector–base junction is reverse biased and so, like any diode, an avalanche effect occurs above a certain voltage and the current rises rapidly. Note that the output current, I_C, now depends on I_B and V_{CE}. Apart from this non-linearity, the current is noisy and probably flows at the least perfect part of the junction. This will cause a hot-spot and probably an early failure of the device.

A good transistor is one with a low leakage current, high V_{CE} before avalanche, evenly spaced lines on the output characteristic and a low knee voltage. Data are shown in appendix B of a typical cheap npn transistor, type BC107. Individual samples may differ considerably from the maker's estimated mean of his product. In particular, the tolerance on the common-emitter current gain, β, or h_{fe} may be as high as -50 to $+100$ per cent on a cheap unselected device.

3.5 Data for a typical npn transistor

The complete maker's data for any device may appear lengthy and confusing when a student first meets them. Appendix B shows some data extracted from several pages compiled by the maker. Further reading will be needed for a student to become familiar with all the terms used. However, the data will be used to calculate real performance figures in the examples of amplifiers that follow.

The four main sections discernible in a maker's data are:

(*a*) the safe maximum ratings allowed for the device,

(*b*) the normal operating voltages, currents and gains,

(*c*) some parameters for a model of the transistor, and

(*d*) typical graphical characteristics.

The sections for (*a*), (*c*) and (*d*) are headed 'Ratings', '*h*-parameters' and 'Typical characteristics' respectively in appendix B. The section for (*b*) usually headed 'Electrical characteristics' can also be found in detailed data from manufacturers.

Certain npn and pnp transistors are available with closely matching characteristics (with polarities reversed). These are called complementary pairs of transistors. Some examples are: npn type – BC182, and corresponding pnp type – BC212; or npn type – 2N3903 and corresponding pnp type 2N3905.

3.6 Ratings and selection of operating point

The operating point, sometimes called the working or quiescent point, of the device should be in a *safe* area of its characteristic and in most cases in an area where the characteristics are evenly spaced.

The maker's data for any device usually give clearly

(*a*) Maximum collector current, $I_{C\,max}$.

(*b*) Maximum collector voltage (above which an avalanche may start).

(*c*) Maximum power dissipation, P_{max}. This may be given by the makers

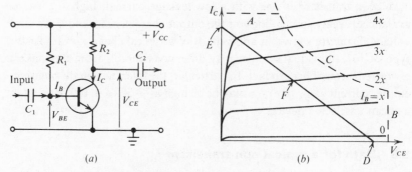

Fig. 3.6. (*a*) The circuit of a simple amplifier and (*b*) the limitation of the operating point shown on the output characteristic.

or it may have to be calculated for the larger devices which can be mounted on a heatsink from

$$P_{max} = (T_J - T_A)/(\theta_1 + \theta_2 + \ldots), \tag{3.6}$$

where T_J is the safe junction temperature given by the maker (usually, 85 °C for germanium and 150 °C for silicon), T_A is the ambient air temperature around the device (it may be enclosed with other power dissipating devices so T_A may be 40 °C rather than the 20 °C of a warm room), and θ_1, θ_2, etc. are the thermal resistances (in degC/watt) of all the elements in the heat flow path between the junction and air. Thus θ_1 will be the thermal resistance between the junction and the transistor mounting base which is given by the maker and depends on the case size and type (about 4 degC/watt for a power transistor in a TO3 case, and 200 degC/watt for the BC107). θ_2 may be the thermal resistance of the insulating washer between the device and the heatsink, θ_3 will be the thermal resistance of the heatsink to air path. (For example, θ_3 is ~ 10 degC/watt for $\frac{1}{8}$ inch thick dull painted aluminium of area 60 square centimetres or 10 square inches.)

Once P_{max} has been obtained, it must equal the electrical power input to the transistor which for the amplifier circuit shown in fig. 3.6(*a*) is given by:

$$P_{max} = V_{CE}I_C + V_{BE}I_B. \tag{3.7}$$

Since $I_B \ll I_C$ and $V_{BE} < V_{CE}$, the second term can be neglected so that a given power dissipation is given by the product of collector current and voltage and can be drawn as a rectangular hyperbola on the output characteristic of the transistor. This clearly limits the usable part of the output characteristic on fig. 3.6(*b*) to the region avoiding the lines of: *A*, maximum safe collector current; *B*, maximum safe collector voltage; and *C*, maximum power dissipation.

When the transistor is to be used as an amplifier, the collector current

84

flows through some load resistor, R_2, in order to develop an output voltage. If the supply voltage is V_{CC}, then the collector current I_C flowing through R_2 gives

$$V_{CC} = I_C R_2 + V_{CE}. \tag{3.8}$$

The two terms, V_{CE} and I_C, are the variables of the output characteristic of the transistor and differentiating (3.8) gives a linear relation because $dV_{CE}/dI_C = -R_2$.

The two points where this line intersects the axes are given by $I_C = 0$, where (3.8) gives $V_{CE} = V_{CC}$; and $V_{CE} = 0$, where $I_C = V_{CC}/R_2$. These are shown as D and E on the graph and the line joining D to E defines the *only combinations of V_{CE} and I_C that are possible once V_{CC} and R_2 have been selected.* This is called the load line.

It is clearly possible to keep the device operating safely by keeping the load line D to E from crossing the boundaries A, B and C. The next problem is that of deciding which combination of collector voltage V_{CE} and collector current I_C to choose out of all those possible.

To obtain a large output voltage swing without distortion, the midpoint F of the load line should be the mean operating point. For no signal input, the base current, $I_B = 2x$ should be drawn through the resistor R_1. An applied signal that varies this by $\pm 2x$ will give an output voltage swing nearly down to zero and up to V_{CC}. The base-biasing resistor, R_1, is calculated from the equation,

$$V_{CC} = R_1 I_B + V_{BE}. \tag{3.9}$$

If some lower collector current is required, e.g. in battery operated equipment, then a lower base current $I_B = x$ say, could be defined by increasing R_1. Note that it would not be possible to get a large output voltage swing unless R_2 was increased to bring the operating point back to the centre of the load line.

This simple amplifier should only be used in cases where the current gain of the transistor is known or can be measured and where it is known that the circuit will not be subjected to wide temperature changes. In other cases, for example if amplifiers are to be mass produced and then are to be mounted in, say, cars where the temperature in the engine compartment is high or in aeroplanes where the temperature outside the pressurised cabin gets very low in flight, then the circuits of the next section should be used.

†3.7 Operating point stability

The base-biasing circuit in fig. 3.6(*a*) only sets up the circuit for a base current I_B of a certain value. However, the static current gain, I_C/I_B, is

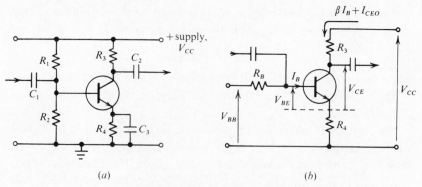

Fig. 3.7. (*a*) A bias circuit giving good operating point stability and (*b*) its equivalent circuit.

very variable from device to device and so a collector current may be realised that is very different from that suggested by point F in fig. 3.6(*b*). In addition the leakage current I_{CEO} which varies sharply with temperature, also flows through the load resistor and will cause the operating point to change with temperature. Thus an improved bias network is wanted for amplifiers which are to be made from untested devices and which will work over a wide temperature range.

Consider the circuit shown in fig. 3.7(*a*). The base of the transistor now draws current from two resistors, R_1 and R_2, connected between the supply lines of the circuit. Current in the output circuit is drawn to the collector through a load resistor, R_3, and flows out of the emitter to the other supply line through R_4. We want to devise a relation between the steady voltages and currents in the circuit which settle the transistor at a well defined working point. In later sections we will consider how small signals, which may be pulses or sine waves, change the base current and hence give a larger signal output from the circuit.

The circuit can be most easily analysed by using Thévenin's theorem to replace the resistors R_1 and R_2 by a single resistance R_B and a new potential source V_{BB}, where

$$R_B = R_1 R_2/(R_1 + R_2), \tag{3.10}$$

$$V_{BB} = V_{CC} R_2/(R_1 + R_2). \tag{3.11}$$

(These assume that the supply V_{CC} has negligible internal resistance.) The circuit of fig. 3.7(*a*) is equivalent to that of fig. 3.7(*b*) and for a steady base current I_B, and collector current I_C, the current out of the emitter flowing through R_4 is $I_B + I_C$. Hence the equation relating the voltages round the base loop is

$$V_{BB} = R_B I_B + V_{BE} + R_4(I_B + I_C) \tag{3.12}$$

and the collector current from (3.2) is

$$I_C = \beta I_B + I_{CEO}.$$

We wish to relate the collector current I_C to the various voltages and resistor values, so re-arranging (3.12) and eliminating I_B gives:

$$V_{BB} - V_{BE} = (R_B + R_4) I_B + R_4 I_C$$
$$= (R_B + R_4)(I_C - I_{CEO})/\beta + R_4 I_C$$
$$= [I_C(R_B + (1+\beta) R_4)/\beta] - [I_{CEO}(R_B + R_4)/\beta],$$

so
$$I_C = \frac{\beta(V_{BB} - V_{BE})}{R_B + (1+\beta) R_4} + \frac{R_B + R_4}{R_B + (1+\beta) R_4} I_{CEO}. \tag{3.13}$$

In a full treatment of the operating point stability, I_C is differentiated with respect to β and with respect to I_{CEO} to get expressions for $\partial I_C/\partial \beta$ and $\partial I_C/\partial I_{CEO}$ which are stability factors for the circuit. This topic is too advanced to be pursued in this introductory text. It is enough to note here that if the emitter resistor $R_4 = 0$ as in the circuit of fig. 3.6(*a*), (3.13) gives

$$I_C = \beta(V_{BB} - V_{BE})/R_B + I_{CEO}. \tag{3.14}$$

Thus the spread in manufacture from one device to another and the variation of the leakage current, I_{CEO}, with temperature will reflect *directly* in an uncertainty in the collector current, I_C.

However, (3.13) shows that if we can include a resistor R_4 such that $\beta R_4 \gg R_B + R_4$: the equation gives the collector current as

$$I_C \approx \frac{V_{BB} - V_{BE}}{R_4}. \tag{3.15}$$

Note that β, the ill-defined current gain, has disappeared from the expression for I_C. All that is required is that it should be high enough to ensure that $\beta R_4 \gg R_B + R_4$. Typical figures of $R_4 = 1$ kΩ and $\beta = 200$ mean that $R_B = 10$ kΩ would allow this approximation to be made. In the expression (3.15), V_{BE} is not known exactly but, as explained earlier, it is the voltage across a junction in conduction which is 0.5 to 0.9 volts for a silicon device. This uncertainty can be made negligible by choosing V_{BB} to be considerably higher than V_{BE}. Consider a device to be set up with a collector current of 5.0 mA; then with $R_4 = 1$ kΩ, V_{BB} of about 5.6 V would be needed to define I_C to within 10 per cent. With the requirement that $R_B = 10$ kΩ, (3.10) and (3.11) allow R_1 and R_2 to be chosen.

Many of the more detailed textbooks calculate the actual changes of collector current. However for the design method just described, the collector current should change by only about 10 per cent for most foreseeable

87

changes in β, V_{BE} and temperature. This is quite acceptable for most circuit designs. The lower leakage current of silicon also makes it the preferred device in most designs even when the higher value of V_{BE} affects the definition of an exact collector current.

3.8 Graphical estimates of circuit gain

The following method gives a quick estimate of the gain of a simple amplifier.

A common-emitter amplifier with the supply voltage V_{CC} joined to the collector through a load resistor R_L will have a load line defining its working points such as that drawn in fig. 3.8(a). Assume that the circuit used to bias the transistor is known to define the operating point to be in the region O shown on the diagram. A small increase in the base current, i_b, equal to the interval between the lines of constant base current on the output characteristic will cause the device operating point to change from O to P, where P *must still be on the load line*. The resulting change in collector voltage, v_{ce}, can be measured from the characteristic. Note that the collector voltage has decreased, i.e. P is nearer the left-hand axis than O, and so v_{ce} will be a negative quantity.

The base current needed for the steady bias at the point O can be put on the input characteristic of the transistor (fig. 3.8(b)) relating the base voltage to the base current. The *same* small rise in base current, i_b, that was considered on the output characteristic is put on the input characteristic and the intercepts on the characteristic give the required voltage change, v_{be}. So voltage gain $= v_{ce}/v_{be}$, and is the ratio of the measured intercepts on the graphs. For accuracy, the changes v_{ce} and i_b must not disturb the device far from the operating point and this conflicts with practical difficulties of measuring small lengths. When larger disturbances are considered, different values of gain will be obtained and the method shows well the non-linearity between gain and input voltage of the circuit for large signal inputs. It is the characteristic at the input, the base, that largely gives rise to this loss of linearity. The output voltage changes v_{ce} are reasonably linearly related to changes of base current, i_b, as shown by the nearly equal intercepts between the lines of base current and the load line on fig. 3.8(a). Thus an amplifier which is required to give a large output voltage swing will work with much less distortion if it is current driven; i.e. the input then should be represented by a *current* source and the relations in chapter 1 concerning ideal current coupling should be considered.

There are many variations on the load line construction to take into

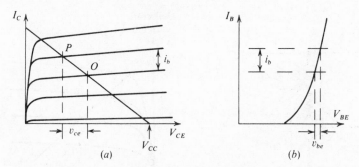

Fig. 3.8. Transistor output characteristic (*a*) and input characteristic (*b*) showing the construction needed for voltage gain estimation.

account the bias resistors in the circuit, the loading by a following stage, and reactive loads. Details of these can be found in the more comprehensive textbooks. However as an illustration, one example is described which shows the principle of the use of DC and AC load-lines; that is the transformer-coupled load.

Fig. 3.9(*a*) shows a bipolar transistor operating from a supply of voltage V_{CC} with the load Z_L connected by a transformer between supply and collector. We wish to relate $V_{CE} + v_{ce}$ to $I_C + i_c$ (the steady component plus the small-signal variations of output voltage and current) to show how the device is kept in the safe area of operation, and to check whether the relation between output and input can become grossly nonlinear for any reason.

The superposition theorem says that one can consider any circuit to be operating with one source at a time and determine the currents that flow due to that source. Then in the presence of all the sources, the current will be the sum of all the individual currents. So consider the stage to be connected to the steady supply V_{CC} but with *no* signal disturbances i_b applied to the base. Then we have

$$V_{CC} = I_C r + V_{CE}, \tag{3.16}$$

where r is the DC resistance of the transformer primary winding. If r is small, a few ohms is typical for a low-loss transformer, then V_{CE} tends to V_{CC}, the supply voltage. The DC load line relating V_{CE} and I_C on the output characteristic of the device is shown in fig. 3.9(*b*). The device voltage is equal to the supply voltage and is independent of the current. A safe current will have to be set by selecting a safe steady base current I_B to flow from the input circuit. Say that this corresponds to $I_B = 2x$, thus a *DC operating point* at O is defined. This clearly is safely under the allowable power dissipation curve.

89

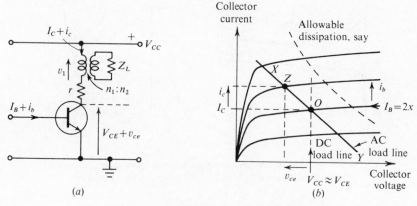

Fig. 3.9. (*a*) Transistor with transformer-coupled load.
(*b*) Output characteristic with load lines.

The question is what will happen when a signal or an AC input, i_b, is applied to the base? The superposition theorem now requires that we calculate i_c and v_{ce}, the resulting signals in the collector circuit. It might appear that if the device continued with the same load line as at DC, then the working point could be disturbed up from O and high collector currents flow that may exceed the device dissipation. However a transformer, which has negligible loss in the winding resistance and in the iron core, has the following simple relations between the voltages and currents with perfectly coupled windings of n_1 and n_2 turns respectively:

$$e_1/n_1 = e_2/n_2 \quad \text{and} \quad i_1 n_1 = i_2 n_2. \tag{3.17}$$

In this circuit, any current in the transformer secondary which is connected to the load, Z_L, is related to the e.m.f., e_2, by

$$e_2 = i_2 Z_L. \tag{3.18}$$

Thus the relation between e_1 and i_1 which applies to the primary of the transformer gives us the impedance looking into the device where

$$\frac{e_1}{i_1} = \frac{e_2 n_1/n_2}{i_2 n_2/n_1} = Z_L \left(\frac{n_1}{n_2}\right)^2.$$

But $v_{ce} + v_1 = 0$ if the supply is decoupled by a capacitor from the V_{CC} line to ground or if it is of low internal impedance so no signals can be developed there; and $i_c = i_1$, the current in the transformer primary. So the relation between signal voltage developed at the collector, v_{ce}, and the current into the collector, i_c, will be given by

$$\frac{v_{ce}}{i_c} = -\frac{v_1}{i_1} = -Z_L \times (\text{Primary:secondary turns ratio})^2. \tag{3.19}$$

This is a linear relation and the locus of operating points as the device is disturbed from the steady operating point O due to signals will be along XOY, the AC load line in fig. 3.9(b). Thus the circuit with base current $I_B + i_b$ will operate at the point Z where the collector current is $I_C + i_c$ and the collector voltage is $V_{CE} + v_{ce}$. At this instant the positive signal current i_c makes v_{ce} negative and hence the transformer voltage v_1 positive.

Equation (3.19) gives the slope of XOY as $-[(\text{turns ratio})^2 \times Z_L]^{-1}$. Since we can order or wind a transformer to have the turns ratio that we require, what would be best for a given transistor characteristic and load Z_L? A first thought may be that we want the load to appear equal to the output resistance that the circuit appears to have; this is the criterion in chapter 1 for best power matching. It is a deficiency of the graphical way of circuit investigation that we have no idea yet what R_o, the output resistance of the device is. Although we can calculate it later in this chapter, and chapter 5 on negative feedback shows how we can modify it, this form of power matching would give rise to an AC load line with a very flat slope so that the circuit will not operate linearly.

In practice a compromise is made and lines of various slopes through O are tried so that X is somewhere near the device knee showing that large increments of both voltage and current occur. The load line should also have evenly spaced intercepts with the lines of constant base current on the characteristic. Then a *current* driven power amplifier stage will have the best linearity possible.

The graphical estimates show well the non-linearity of the device for large outputs, and the inaccuracy due to drawing of about 10 per cent is good enough for quick design work.

3.9 Small-signal equivalent circuits for the bipolar transistor

An alternative method of circuit analysis is possible by deriving an equivalent circuit or model for the transistor for *small signal changes* about an operating point. Three such circuits are shown of the many that are quoted by makers and textbooks.

Fig. 3.10(a) shows the transistor which has to have steady voltages V_{BE} and V_{CE} at its terminals to operate in a safe active mode. If a small signal of voltage v_1 is applied to the input, considered here to be the base, we wish to know what changes result at the output, in either v_2, the change in collector voltage, or i_2, the change in collector current. We need a model which allows us to calculate v_2, then v_2/v_1 will be the device voltage gain;

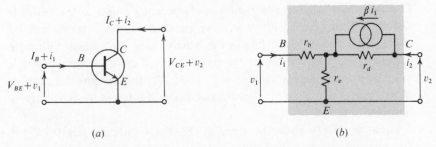

Fig. 3.10. (*a*) The transistor as a three-terminal device and (*b*) the T-network equivalent circuit for small-signal analysis.

and if a current i_1 is found to flow into the terminals of the device, i_2/i_1 will be the device current gain. Also v_1/i_1 will be the impedance that the device presents to previous circuits. All these are valuable performance figures that we will be able to calculate using the equivalent circuit.

†3.9.1 T-Network

This is the simplest circuit to understand as each of the components relates directly to some part of the two-junction physical model for the transistor.

Fig. 3.10(*b*) shows the components in the T-network model. The components connected between the terminals B, C, E (base, collector and emitter) are as follows:

r_e is the low resistance of the forward biased emitter junction (1Ω),

r_d is the high resistance of the reverse biased collector junction (3.3 kΩ),

r_b is the resistance of the thin base layer (750 Ω),

β is the current gain in the common-emitter connection (250).

In brackets are typical figures for the BC107 transistor at an operating point of $V_{CE} = +10$ V, $I_C = 25$ mA, $I_B = 100$ μA and $V_{BE} = +0.7$ V. It is the device whose data are given in appendix B.

The actual values taken by r_d, r_b and β vary between specimens of one given transistor type. For a cheap, unselected device they may be anywhere between half and double the maker's typical figure. r_b can be expected to be larger for high frequency and high gain transistor types because of the thinner base layer these types have.

Since the current gain, β, depends on the doping concentrations in the emitter and base layers of the device and on the thinness of the base layer, and all of these may be variable from one day's production to the next, most makers sort their products into low gain and high gain types. For example, Texas Instruments divide their BC107 series of transistors into

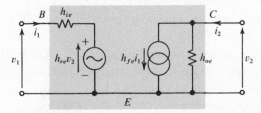

Fig. 3.11. The hybrid-parameter small-signal equivalent circuit for
a common-emitter bipolar transistor.

groups A, B and C where currents gains of 125 to 260, 240 to 500 and
450 to 900 times respectively are expected at $V_{CE} = 5$ V and $I_C = 2$ mA.

The emitter resistance, r_e, however is predictable and depends on the
steady current through a conducting junction. The incremental, slope or
small-signal resistance that a diode appears to have was derived in
chapter 2 and is approximately kT/eI ohms. At room temperature, kT/e
is approximately 1/40 volts, and I is the total current flowing. For the
BC107 operating point already mentioned, $I_C + I_B = 25.1$ mA, thus

$$r_e = \frac{kT}{eI} = \frac{1}{40} \times \frac{10^3}{25.1} \approx 1\Omega. \tag{3.20}$$

This is exactly the figure quoted for the device.

For rough calculations, this equivalent circuit is sometimes simplified
by assuming r_d is a high impedance compared to other paths in the circuit
and by assuming that r_b is a low resistance. However it is better to be
capable of writing expressions for the performance of a circuit without
assumptions of this type. Small terms can later be ignored when actual
component values are used; particularly as a wide spread in current gain is
to be expected from all but the more expensive selected devices.

3.9.2 Hybrid-parameter network

The h-parameters are those most widely quoted in manufacturers' literature
at present. The parameter values come directly from the input and output
characteristics of the transistor. This model is shown in fig. 3.11 for the
common-emitter circuit.

The model contains an impedance, both a voltage and a current generator
and a shunting path at the output, h_{oe} which has units of admittance. It is
the mixture of units of the components of this model that gives the model
its name. At first sight, this circuit may appear more complicated than
the T-network but both contain only four components. These are defined
below and typical figures are given (again for the BC107 transistor at

low frequency and at operating point $V_{CE} = +10$ V, $I_C = 25$ mA, $I_B = 100\mu$A, $V_{BE} = +0.7$ V).

h_{ie} is the base circuit resistance, $|\partial V_{BE}/\partial I_B|_{V_{CE}}$ (1000 Ω),

h_{re} is the reverse voltage transfer ratio, $|\partial V_{BE}/\partial V_{CE}|_{I_B}$ (3×10^{-4}),

h_{fe} is the forward current gain, $|\partial I_C/\partial I_B|_{V_{CE}}$ (250),

h_{oe} is the output admittance, $|\partial I_C/\partial V_{CE}|_{I_B}$ (300 μS).

Each of the components has got a second subscript '*e*' which denotes that the common-emitter connection is specified. An identical circuit can be used for common-base and common-collector connection of the transistor but with the subscript *b* or *c* respectively instead of *e*. (The parameters then have different values.) The first subscript is descriptive in that h_{ie} refers to the *input* path, h_{re} refers to a generator in the input circuit dependent on some output quantity, namely v_2; that is, it is a *reverse* effect from output to input of the device. h_{fe} represents the *forward* gain mechanism of the transistor and h_{oe} is a component solely across the *output* terminals.

In the equivalent circuit, fig. 3.11, the left-hand part has to represent the device looking into the base terminal, *B*. It is of course a conducting junction whose typical characteristics were shown in fig. 3.4(*b*). The slope of the characteristic gives 1/resistance looking into the junction and the dependence on collector voltage is shown by the separation of the characteristics. These have been described fully in the notes referring to fig. 3.4(*b*) and give rise, of course, to the two parameters of the circuit. h_{re} is very small and is often neglected in approximate analysis. h_{ie}, the input resistance, can be calculated from $1/(40I_B)$ since in this model the path is carrying *only* the base current, I_B. For a junction carrying 100 μA, this equation gives a value of 250 Ω for input resistance. The discrepancy between this and the maker's figure of 1000 Ω is due to the resistance in the thin base layer. (We know that this resistance was $r_b = 750$ Ω in the T-network so this agrees well.) A check could also be made that the slope of the characteristic in the BC107 data in appendix B is of this order; it will only be approximate as the device data are often plotted to a very small scale, perhaps to conceal their imprecise nature.

In the equivalent circuit, fig. 3.11, the right-hand part has to represent the transistor looking in at the collector terminal, *C*. This is a reverse biased junction so that the current change that takes place by changing only the output voltage can be expected to be small. This change is taken into account by h_{oe} which can be expected to be a low conductance (or $1/h_{oe}$ is a high resistance). The main current that flows at the collector terminal is dependent on that flowing into the base terminal, i.e. i_1, and is taken into account by the generator $h_{fe}i_1$ where h_{fe} is the current gain and is nearly the

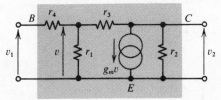

Small-signal equivalent circuits

Fig. 3.12. The hybrid-π network representing a bipolar transistor.

same as the β of the T-network. The derivation of these two parameters from the output characteristics of the transistor, fig. 3.5, has been more fully described in the notes referring to that figure.

The values given for h_{fe} and h_{oe} can be checked from the BC107 data in appendix B. The change in collector current should be of the order of 250 times the change in base current in the region of the $V_{CE} = +10$ V, $I_C = 25$ mA working point. The reader may check this on the graph which also happens to have very evenly spaced characteristics which shows that the current gain is not widely different at other working points. Also the slope of the lines of constant base current near the working point should be of the order of 0.3 mA change in I_C for a 10 V change in V_{CE}; these figures are obtained because the output admittance parameter is 0.3 mS. For approximate calculations, this low admittance, h_{oe}, is sometimes ignored, i.e. replaced by an open circuit.

†3.9.3 Hybrid-π network

This equivalent circuit is widely used in American textbooks. It is important in that, by adding capacitors to the model, the transistor's performance changes at high frequency can also be calculated with reasonable accuracy (see the end of §3.11).

Fig. 3.12 shows the hybrid-π equivalent circuit. The parameter giving the device gain is represented as a forward conductance, g_m, and so the model can be used with similar circuit equations as the field-effect transistor. There are five components in the circuit where:

r_4 is the resistance of the thin base-layer (750 Ω),

r_1 is the resistance of the conducting base–emitter junction (250 Ω),

r_3 accounts for the feedback in the transistor (and is equivalent in effect to h_{re} in the hybrid model and r_e in the T-network) (1 MΩ),

r_2 is the output resistance (3.3 kΩ),

g_m is the forward conductance (approximately $40I_C$) (1 S).

In brackets are typical figures for the BC107 transistor at an operating point of $V_{CE} = +10$ V, $I_C = 25$ mA, $I_B = 100$ μA and $V_{BE} = +0.7$ V. By comparing the typical figures of this circuit with the T-network and

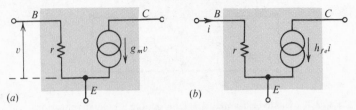

Fig. 3.13. Approximate bipolar transistor equivalent circuits.
$r \approx 1/40I_B \,\Omega, \, g_m = 40I_C$ S.

hybrid-parameter network the following approximate relations are helpful in converting data,

T-network	Hybrid	Hybrid-π	
$r_b + \beta r_e$ =	h_{ie} =	$r_1 + r_4$	(3.21)
r_e/r_d =	h_{re} =	r_1/r_3	(3.22)
β =	h_{fe} =	$g_m r_1$	(3.23)
r_d =	$1/h_{oe}$ =	r_2	(3.24)

It is often found that the manufacturers do not give typical characteristics for a cheap device in which there might be a considerable spread in the current gain. Then the relations help convert *any* data that *are* known into an approximate equivalent circuit.

Knowing only the common-emitter current gain, h_{fe} or β for a transistor, the circuits shown in fig. 3.13 can be used. The reader can derive the elements himself for the two circuits by simplifying the circuits of figs. 3.11 and 3.12. In these figures, if we ignore the various paths away from the output generators, we are left with two generators only and these must be identical, so

$$h_{fe}i = g_m v, \qquad (3.25)$$

so

$$g_m = h_{fe}\frac{i}{v} = \frac{h_{fe}}{r} = h_{fe}.40I_B$$

or approximately $\qquad 40I_C$ S.

The simplifications that $i/v = r$ and $h_{fe}I_B = I_C$ are approximate, but where a calculation is desired to estimate only the order of magnitude of performance, this simplification is often used.

3.10 Calculation of amplifier performance

It is not possible in a short book to give examples of the use of all the equivalent circuits. Books using all the various networks are listed in appendix A. The following analyses only use the *h*-parameters.

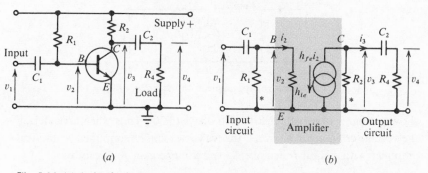

(a)   *(b)*

Fig. 3.14. (*a*) A simple single stage amplifier and (*b*) its equivalent circuit for calculation of performance. * Denotes supply+assumed to be low impedance to ground.

The method of analysis in each case is to replace the transistor by its equivalent circuit. Thus the transistor in fig. 3.14(*a*) with its three leads marked E, B and C, is replaced by its equivalent circuit between the same leads in circuit (*b*) and is included within the grey tinted area. Here the simplified h-parameter model is shown in which h_{re} and h_{oe} are neglected.

Now looking out from the terminals E, B, C, in fig. 3.14(*a*), all the current paths connected to each of the terminals are added to complete the circuit. Thus looking out from the base, there are two paths – through C_1 to the source v_1 and through R_1 to the positive supply. The latter is considered to have a low impedance to the signal common line so one end of R_1 is shown connected to that line in fig. 3.14(*b*). The circuit connected to the device output side is similarly inserted. We then develop expressions relating v_2, v_3, i_2 and i_3. Then v_1 and v_4 can be calculated when the source and load are fully defined by using the equations of chapter 1 relevant to passive coupling circuits.

Remember that the following are for the simplified equivalent circuit of the device (h_{oe} and h_{re} neglected),

Voltages at the input give
$$v_2 = i_2 h_{ie}.$$

Currents at the collector node give
$$i_3 + v_3/R_2 + h_{fe} i_2 = 0.$$
Eliminating i_2,
$$v_3 = -v_2 h_{fe} R_2/h_{ie} - i_3 R_2 = v_2 A_v - i_3 R_2, \qquad (3.26)$$

or $v_3 = v_2$ (no-load voltage gain) $-i_3$ (output resistance).

The amplifier can thus be represented by an ideal voltage amplifier,

97

The bipolar transistor

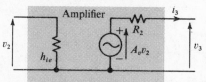

Fig. 3.15. Equivalent voltage amplifier to the circuit of fig. 3.14(b).

shown in fig. 3.15, which clearly also obeys (3.26). Thus when considering the voltage coupling in a circuit, the transistor can be simplified to an ideal amplifier with output resistance R_2 and voltage gain A_v given by

$$A_v = -h_{fe} R_2/h_{ie}, \tag{3.27}$$

where R_2 is the collector resistor and the sign represents $180°$ phase shift between input and output. This gain will be the maximum possible from the stage if the input and output coupling circuits are nearly perfect and only a very little *voltage* attenuation takes place. Equation (1.1) showed that this was only true when the load on the source was much greater than the internal resistance of the source. In this circuit, nearly the opposite is true, h_{ie} is low and R_2 may often be high. Now a circuit of low input resistance and high output resistance more nearly characterises an ideal current amplifier. Thus it is more profitable often to consider a bipolar circuit as a current amplifier.

In fig. 3.14(b), the circuit in the tinted area is a current amplifier with gain, $A_i = -h_{fe}$, input resistance $= h_{ie}$ and output resistance $= R_2$. By considering *current* coupling at the input and output circuits, the overall circuit gain can be calculated. This is done in the example at the end of the chapter.

An idea of the order of magnitude of the gain is useful. Using the data for the BC107 transistor in appendix B, and assuming that the circuit is such that the load $R_2 = 400 \, \Omega$, and the device has $h_{ie} = 1000 \, \Omega$ and $h_{fe} = 250$, the amplifier will have a voltage gain $A_v = -100$ (or $A_i = -250$), input resistance $= 1000 \, \Omega$ and output resistance $= 400 \, \Omega$.

It is also interesting to check by what factor these results are inaccurate because h_{oe} and h_{re} have been ignored. Using the typical figures of 300 μS and 3×10^{-4} respectively (see §3.9.2) the exact equations for the performance of the circuit give: $A_v = -90$, $A_i = -223$, input resistance $= 973 \, \Omega$ and output resistance $\approx 350 \, \Omega$. Clearly errors of $\sim 10\%$ that come from simplifying the equivalent circuit are not important when the transistor data itself are not guaranteed to be very accurate.

98

3.11 Frequency response

At low frequencies, the effect of C_1 and C_2 can be obtained by analysis of the input and output circuits. At high frequencies, the value of the parameters in the transistor model are no longer real; say, h_{fe} becomes $100 - j40$ at 3 MHz instead of $250 + j0$ at low frequencies. This can be put into the expression for gain which will then be calculated as a complex number. Converting to the alternative magnitude and phase angle form, we can see by how much the gain has dropped and what phase shift the circuit then gives.

The capacities associated with the junction can be shown on the hybrid-π model and added paths taking currents that are frequency dependent are inserted in the equivalent circuit, say such as that in fig. 3.14(b). Now new expressions relating the voltages and currents of the circuit can be written.

Or we can use the relationships already obtained in chapter 2. If we know the collector to base junction capacity C_{cb} in the device, then the collector node sees an added current flowing out and a capacitor C_{cb} appears to shunt the output. Equation (1.21) showed how to work out the high frequency at which such a capacitor and other shunting paths would decrease the output. It depended on the total parallel resistance. Thus for the circuit shown in fig. 3.14(b), the resistances at the output are R_2 and R_4 and so a collector capacity C_{cb} would define a turnover frequency ω_2, where

$$\frac{1}{\omega_2 C_{cb}} = \frac{R_2 R_4}{R_2 + R_4}. \tag{3.28}$$

However at the input of the device, extra current will flow into the base because of C_{cb} connected to the base, and from § 2.15 we know that this capacity appears as $(1 - A_v) C_{cb}$ because of the voltage gain A_v of the circuit. Again the turnover frequency ω_3 can be estimated as that when the capacitor impedance is equal to the resistance of all other paths in parallel between the base and signal zero of fig. 3.14(b). Thus if R_s is the resistance of the source of e.m.f. v_1, writing that the *admittances* are equal,

$$\omega_3 (1 - A_v) C_{cb} = \frac{1}{R_s} + \frac{1}{R_1} + \frac{1}{h_{ie}}. \tag{3.29}$$

When the component values are known, then ω_2 and ω_3 can be calculated. Fig. 3.16 shows the form of frequency response that the circuit will have. At a frequency lower than both ω_2 and ω_3 the gain will be largely constant.

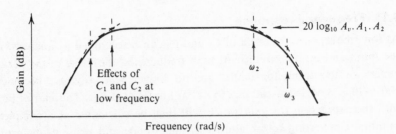

Fig. 3.16. Typical response of the bipolar amplifier.

On fig. 3.16, this gain has been shown as $20 \log_{10} A_v A_1 A_2$ decibels (where A_1 and A_2 are the input and output coupling gains) and this assumes that the circuit is loaded by a resistance equal to its input resistance; i.e. it could be one of a series of identical amplifiers in cascade. Between ω_2 and ω_3 the gain falls as frequency increases at 6 decibels per octave (or 20 decibels per decade). At a frequency higher than ω_3, the gain falls more swiftly as frequency rises, (12 decibels per octave or 40 decibels/decade). The input and output circuits also have stray shunting capacities in them and so C_s should be added as a term in both (3.28) and (3.29). It means that there is more phase shift at high frequencies than might at first be thought. Chapter 5 explains the implications of phase shift at high frequencies, and how feedback in amplifiers of two or more stages can give instability.

†3.12 Emitter follower performance

In §3.10, the performance of the common-emitter connection for the bipolar transistor was calculated. The simple circuit represented a good current amplifier with low input resistance. To make amplifiers which will respond linearly to the *voltages* from a source, three circuits are common: an emitter follower stage, which is described here; a multistage amplifier with a field-effect transistor as the first stage, which is described in §3.13; and a differential amplifier, which is described in chapter 4.

The emitter follower amplifier (also called the common-collector connection) is similar to the cathode follower valve circuit or source follower f.e.t. circuit and it is shown in fig. 3.17.

The principle of superposition allows us to consider separately the circuit first for the application of the steady supply voltage and then for a small signal v_1 applied to the circuit. The resistors R_1 and R_2 define the steady working point of the transistor. By assuming for the circuit the

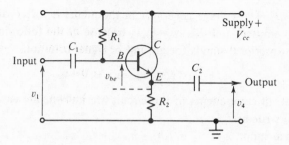

Fig. 3.17. Emitter follower circuit arrangement. Clearly $v_1 = v_{be} + v_4$, so voltage gain $= v_4/v_1 \approx 1$.

desired values, V_{CE}, I_C, V_{BE} and I_B, that are required for the operating point, and with the supply voltage V_{CC}, the resistors are,

$$R_1 = \frac{V_{CE} - V_{BE}}{I_B} \quad \text{and} \quad R_2 = \frac{V_{CC} - V_{CE}}{I_C + I_B}. \qquad (3.30)$$

The capacitors C_1 and C_2 are assumed to be perfect in blocking the steady voltages of the circuits providing the input and receiving the output respectively.

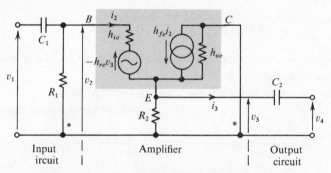

Fig. 3.18. Equivalent circuit of the emitter follower using *h*-parameters.
* Denotes supply + assumed to be low impedance to ground.

For small signals, by replacing the transistor by the *h*-parameter equivalent circuit in the grey tinted area, the emitter follower is equivalent to the circuit in fig. 3.18. Again the positive supply line is assumed to be decoupled so signals are not developed there. Thus R_1 and the collector of the transistor are shown joined to the signal common line.

We desire to develop relations between v_2, v_3, i_2 and i_3. The relationships v_2/v_1 and v_4/v_3 depend on the details of the input and output circuits respectively, and can be calculated from the theory of chapter 1. Because the signal current into the transistor base is i_2, the current generator in the

101

model is $h_{fe}i_2$. The voltage generator in the model is $h_{re} \times$ (voltage from collector to emitter) and this is $-h_{re}v_3$ because of the following relation where the subscript 0 stands for ground or signal common,

$$v_{ce} = v_c - v_e = v_{c0} - v_{e0} = 0 - v_3. \tag{3.31}$$

Now that all components in the circuit are known, the easiest circuit relations to write are:

Voltages at input, $\qquad v_2 = i_2 h_{ie} - h_{re}v_3 + v_3. \tag{3.32}$

Currents at emitter node,

$$i_2 + h_{fe}i_2 + (-v_3)h_{oe} = v_3 G_2 + i_3, \tag{3.33}$$

where G_2 is the conductance of R_2, so

$$G_2 = \frac{1}{R_2} \text{S (siemens),} \tag{3.34}$$

where R_2 is in ohms.

From these two equations, we can eliminate any one unknown. If we first eliminate i_2, the input current, we are left with the output voltage v_3 in terms of the input voltage v_2 and output current i_3 as

$$v_3 = \frac{v_2(1 + h_{fe}) - i_3 h_{ie}}{h_{ie}(G_2 + h_{oe}) + (1 + h_{fe})(1 - h_{re})}. \tag{3.35}$$

Now this can be compared to the expression for the output of any amplifier which can be written,

Output voltage = input voltage × no-load voltage gain

$\qquad\qquad\qquad\qquad\qquad$ − output current × output resistance.

No-load voltage gain $= \dfrac{1 + h_{fe}}{h_{ie}(G_2 + h_{oe}) + (1 + h_{fe})(1 - h_{re})}$

or approximately $\qquad\qquad = \dfrac{h_{fe}}{h_{ie}(G_2 + h_{oe}) + h_{fe}}, \tag{3.36}$

as typically $h_{fe} \gg 1$ and $h_{re} \ll 1$. In deriving expressions such as (3.35) or (3.36), it is good practice to check quickly that the dimensions of the expression are correct and so no slips have been made. Also it is apparent that the value for the gain given by (3.36) can never exceed unity. By considering the signal voltages shown in fig. 3.17, this again is to be expected and is a check on the correct derivation of the expression.

The output resistance of the amplifier is also obtained from (3.35) as

$$\text{Output resistance} = \frac{h_{ie}}{h_{ie}(G_2 + h_{oe}) + (1 + h_{fe})(1 - h_{re})}. \tag{3.37}$$

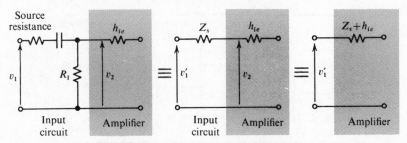

Fig. 3.19. Equivalent input circuits.

Again assuming $h_{re} \ll 1$ and taking the reciprocal of the expression, we get

$$\text{Output admittance} = G_2 + h_{oe} + \frac{(1+h_{fe})}{h_{ie}}. \tag{3.38}$$

This represents three paths in parallel. The paths through R_2, the emitter resistor, and through h_{oe} in the transistor model are obvious. But the other path is that which gives the circuit its desirable high output admittance (equivalent to *low output resistance*), so the output voltage is little changed if the circuits that follow draw current.

It should be noted that h_{ie} is not the only resistance in the input path in fig. 3.18. To it should be added the equivalent series impedance of the bias and source circuits and this can be calculated using the Thévenin equivalent shown in fig. 3.19. Thus the effect of the input circuit can be taken into account by using the new input voltage v_1', instead of v_2 in (3.35) and by using $Z_s + h_{ie}$ in place of h_{ie} alone in (3.35) and (3.38). In practice the authors have found that engineers fall into fewer errors by making these separate steps of calculation. The magnitudes of v_1' and Z_s can quickly be checked by inspection when numerical values are known. A mathematician may develop an expression double the length of (3.37) which allows for generator impedance, bias resistors, etc., but it is less easy to check quickly that the results obtained from it sound sensible.

Another performance feature of the amplifier that we wish to calculate is its input resistance. If we assume that the load on the amplifier due to following circuits is equivalent to a single component of admittance G_L, then

$$i_3 = v_3 G_L.$$

With (3.32) and (3.33), we now have three equations relating the four unknowns of the circuit. By eliminating v_3 and i_3, we are left with one equation for v_2 in terms of i_2 which gives

$$\text{Input resistance} = \frac{v_2}{i_2} = h_{ie} + \frac{(1-h_{re})(1+h_{fe})}{h_{oe}+G_2+G_L}. \tag{3.39}$$

103

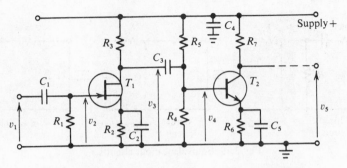

Fig. 3.20. Interstage coupling of a multistage amplifier.

With the simplification that $h_{re} \ll 1$, the amplifier can be seen to have 'transformed up' the total impedance of the emitter to ground paths by a factor of $(1 + h_{fe})$.

Actual performance figures for the emitter follower circuit can be calculated by those doing problem 5 at the end of this chapter. A no-load voltage gain of the order 0.95 to 0.99, an output resistance of a few hundred ohms and an input resistance of the order of a megohm are common. The circuit does not really contribute to the performance of any amplifier by its voltage gain but as an impedance transformer: it is commonly found as the first and last stages of most integrated circuit voltage amplifiers.

†3.13 Coupling between amplifier stages

In practice, we find that amplifying devices are connected in series (or cascade) to give a circuit of gain higher than that we actually require. The gain is then dropped by feedback, which gives us several other benefits, to exactly the amount required for a particular application (see chapter 5).

Fig. 3.20 shows the first two stages of a typical voltage amplifier which uses first a field-effect transistor because of its high input resistance and then bipolar transistors of which only the first is shown. The bias conditions for the two devices shown are derived quite separately and C_3 blocks the steady drain voltage of T_1 from appearing at the base of T_2. This is called resistor–capacitor or 'R–C' coupling. To find the gain, we write equations to relate v_1, the small-signal input voltage to the amplifier, to v_3, the output of the first stage; then to relate v_4 to v_3 and so on.

We know that in choosing R_4 and R_5, low values will give good stability of the working point of T_2 against temperature changes (§3.7). But how low should they be? 10 kΩ was chosen in an example as the parallel resistance of R_4 and R_5 but is this too low? Answers come from this apparently

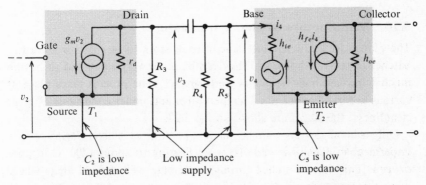

Fig. 3.21. Equivalent circuit to that of fig. 3.20 for small signals.

complicated circuit when we use the small-signal models for the transistors T_1 and T_2. We assume that the frequency of the signals is such that the impedances of the decoupling capacitors, C_2, C_4 and C_6 are low. Then the circuit, for the components affecting the coupling between the transistors, becomes that shown in fig. 3.21. The output of T_1 is normally represented as a current generator, $g_m v_2$, in parallel with the drain resistance, r_d. The input of T_2, using the hybrid parameters, is represented by an impedance h_{ie} and a generator into which we want to maximise the input current i_4. This is because the output of T_2 is a generator dependent on i_4; so the problem is one of getting good *current* coupling. In its simplest terms, we want to get all of the current output of T_1 flowing into T_2; so all the paths parallel with h_{ie} should be much higher in impedance than h_{ie} (see (1.6)). Assume that we knew r_d to be of the order of 30 kΩ: then we would wish R_3, R_4, R_5 to be of the same order and h_{ie} to be one or two kilohms only to get reasonable current coupling. This may sound a very casual approach to circuit design, but it is not profitable for a designer always to analyse accurately the exact current coupling factor, when the gain of the transistors, h_{fe} and g_m, may be between a *half* and *double* the nominal value given by the device makers! The aim should be not to waste gain by poor coupling between the output of one stage into the input of the next.

The alternative to using the *R–C* coupling, that has been described here, is to use direct coupling. Amplifiers without interstage capacitors are called DC (direct coupled) amplifiers. The output voltage of one stage must now provide the right steady bias for the next stage. Such circuits are widely used and are described in texts listed in appendix A.

3.14 Manufacture of planar transistors

The planar technique enables silicon transistors to be mass-produced in batches of several thousand. They can be made very small and so have a much improved high frequency performance. The process takes place at various accurately controlled temperatures and under conditions of great cleanliness: the stages are shown in fig. 3.22.

Only silicon devices can be made like this as silicon forms a thin impermeable insulating oxide layer on heating to about 1200 °C in pure oxygen. This can be etched through to create well defined areas where doping compounds are deposited on to the surface and then diffused into the semiconductor at elevated temperatures.

Fig. 3.22 shows the *cross-section* of one small piece of the silicon wafer to show how the npn layers are built up. The surface dimensions of this one device may be of the order of $\frac{1}{3}$ mm by $\frac{1}{3}$ mm, and identical diffusions will be put on to a pattern of similar areas covering the whole silicon disc with similar devices. The *plan* view of the disc is shown at the top right of fig. 3.22 and the remaining figures show how small the active area of the device is compared to the normal three-lead case that eventually contains it.

The reader who wishes to have more explanation of the production methods must consult other texts.

3.15 Worked example

The transistor shown in fig. 3.23 has $h_{ie} = 1.25\ \Omega$, $h_{fe} = 50$, $h_{oe} = 20$ mS and h_{re} negligible at a working point of $V_{CE} = +12$ V, $I_C = 1.0$ A, $I_B = 20$ mA and $V_{BE} = +0.5$ V.

(a) For a 25 V supply, and if the voltage across $R_4 = 1$ V and the current in $R_2 = 4I_B$, what values are needed for all the resistors?

(b) What is the optimum load and what power is developed in it by 15 mA RMS input current?

(c) What is the stage efficiency and power gain?

(d) The transistor has a thermal resistance from junction to case of 2 deg/watt and is mounted on a heatsink of thermal resistance to ambient air of 5 deg/watt. For ambient air at 40 °C max., is the junction safe at under 150 °C?

Solution. (a) After carefully marking the transistor working point conditions, the supply voltage and the given conditions at R_2 and R_4 on the circuit, the component values can be written down as the voltage across them divided by the current through them.

$$R_3 = \text{collector resistor} = \frac{25 - (12 + 1)}{1.0} = 12\ \Omega$$

(power rating needed, $I^2R = 12$ watts).

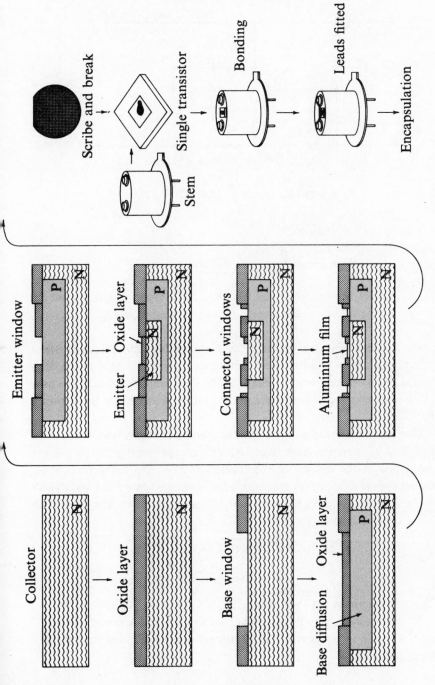

Fig. 3.22. Stages in the manufacture of silicon planar transistors (by courtesy of *Mullard Outlook*).

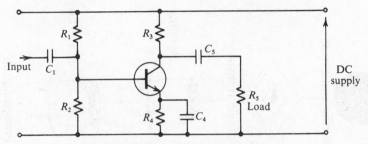

Fig. 3.23. Circuit for worked example.

R_4 = emitter resistor = $1.0/1.02 \approx 1 \ \Omega$.

$$R_2 = \frac{1.0+0.5}{4 \times 0.02} = 18.75 \ \Omega.$$

$$R_1 = \frac{25-(1.0+0.5)}{5 \times 0.02} = 235 \ \Omega.$$

So R_2 and R_1 may be chosen as 18 Ω and 220 Ω types. (1 watt, $\frac{1}{4}$ watt and 3 watt types will be needed for R_4, R_2 and R_1 respectively.)

(b) To determine the optimum load, the equivalent source resistance that the transistor and its circuit appears to present will have to be determined: then the best load is that equal to the source resistance. So using the h-parameter equivalent circuit, the amplifier is redrawn as shown in fig. 3.24.

One has to make the assumptions that the power supply is well decoupled and that C_4 can be made large enough so that negligible signals are developed at the emitter of the transistor. Then the transistor and its circuit appear as a current generator of $h_{fe}i_2$ shunted by h_{oe} in parallel with R_3 feeding the load. As $h_{oe} = 20 \times 10^{-3}$ S and $R_3 = 12 \ \Omega$, this shunt path has a conductance G_0 given by:

$$G_0 = 20 \times 10^{-3} + 1/12 = 103 \times 10^{-3} \text{ S},$$

$$R_0 = 1/G_0 = 9.7 \ \Omega.$$

Thus the optimum load will be one of 9.7 Ω (§1.6).

To calculate the power in this load, the currents in the circuit are calculated when $i_1 = 15$ mA RMS; if R is the resistance of R_1 and R_2 in parallel (approximately 18 Ω),

$$i_2 = \frac{Ri_1}{R+h_{ie}} = \frac{18 \times 15 \times 10^{-3}}{18+1.25} = 14 \times 10^{-3} \text{ A RMS}.$$

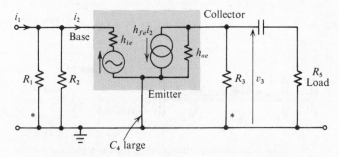

Fig. 3.24. Equivalent circuit to that of fig. 3.23 for small signals.
* Denotes DC supply well decoupled.

Current generator in output, $h_{fe}i_2 = 0.7$ A RMS. Half of this current will go into the load, as the load resistance is equal to that shunting the current generator, so

$$\text{Output power} = (0.35)^2 \times 9.7 = 1.2 \text{ watts.}$$

(c) Fig. 3.25 shows the amplifier as a 'black box' and the input and output powers are marked.

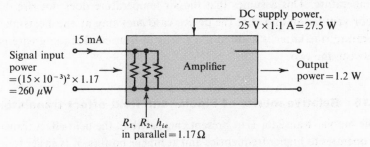

Fig. 3.25. Power inputs and outputs for the amplifier.

$$\text{Efficiency} = \frac{\text{output power}}{\text{total power input}} = \frac{1.2}{27.5 + 0.00026} = 4.4 \text{ per cent.}$$

$$\text{Power gain} = \frac{\text{signal output power}}{\text{signal input power}} = \frac{1.2}{0.00026} = 4600.$$

(d) The thermal resistance of a given path is defined as the temperature difference needed to get a given heat flow rate. The relation can be restated as

$$\text{Temperature difference} = \text{thermal resistance} \times \text{watts dissipated.}$$

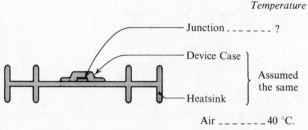

Fig. 3.26. Cooling arrangements for the transistor.

The layout of a transistor on its heatsink is shown in fig. 3.26. First we calculate the heat flow rate as:

$$\text{Power dissipated} = V_{CE} \times I_C + V_{BE} \times I_B = 12 \text{ watts.}$$

Thus the temperature differences are:

$$\text{Heatsink to air} = 5 \times 12 = 60 \text{ degC.}$$

$$\text{Junction to heatsink} = 2 \times 12 = 24 \text{ degC.}$$

So the junction is 84 °C above the ambient air and is at 124 °C, a safe temperature. This assumes that the air temperature does not rise due to poor ventilation and that the device case *does* stay at the heatsink temperature: the latter is assured if a smear of heat conducting grease is put between the two.

3.16 Relative merits of bipolar and field-effect transistors

The bipolar transistor is at present cheaper than the field-effect transistor. It operates to higher frequencies and at higher powers; it is easier to use in integrated circuits and is a very good current amplifier.

The field-effect transistor, with its high input resistance and low noise, is to be preferred for the first stages in amplifying small voltage signals from high impedance sources; such as skin potentials in medical electronics. It is less affected by radiation than the bipolar transistor whose leakage current will rise when in the presence of radiation.

3.17 Summary

Setting up the transistor at a safe operating point by choosing the DC supply, the load resistor and the bias circuit component values *is a quite separate problem* from working out the circuit performance figures such

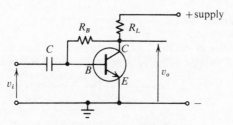

Fig. 3.27. Circuit for problem 1.

as gain, output impedance, etc. from the small-signal equivalent circuit of the device.

Choosing the DC supply and load resistor will define a load line across the device characteristics in exactly the same way as that considered for the field-effect transistor in chapter 2. However the bipolar transistor requires a bias *current* to its base to define the operating point on the load line. If this is provided by a circuit in which the collector current is not sensed, then the wide variations usually found with all but specially-selected expensive transistors will give wide variations in collector currents – and so an imprecise working point results together with possible distortion starting at quite low output voltage swings.

The *h*-parameter model for the bipolar transistor appears more complicated than the model used for the field–effect transistor. However it has been shown to give good estimates of circuit performance even if it is simplified by omitting consideration of h_{re} and h_{oe}. Examples of its use are given for a common-emitter amplifier (§3.10), for a common-collector amplifier, or emitter follower, which has a high input resistance and low output resistance (§3.12) and in a power amplifier (§3.15).

3.18 Problems

1. The BC107, silicon npn transistor has the following typical working point:

$$V_{ce} = +10 \text{ V}, \quad I_c = 25 \text{ mA}, \quad I_b = 100 \,\mu\text{A}, \quad V_{be} = +0.7 \text{ V}.$$

At this working point, its small signal parameters are:

$$h_{ie} = 1000 \,\Omega, \quad h_{re} = 3 \times 10^{-4}, \quad h_{fe} = 250, \quad h_{oe} = 300 \,\mu\text{S}.$$

(*a*) If the supply voltage is $+20$ V, what load resistor and base bias resistor will achieve the given working point in the simple common-emitter amplifier of fig. 3.27?

(*b*) Find the current gain, input resistance, voltage gain and power gain for the amplifier of fig. 3.27. (Ignore the effect of C and of R_B in the circuit.)

(*c*) Graphical data of the BC107 are in appendix B. For a supply of $+20$ V,

111

put a 400 Ω load line across the output characteristic and check that the working point at (*a*) is obtainable.

For an input swing of ± 50 μA about the standing base current of 100 μA, get graphically the voltage gain of the amplifier with a 400 Ω load.

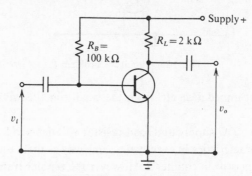

Fig. 3.28. Circuit for problem 2.

2. (*a*) Derive the voltage, current and power gain relationships and then find the voltage gain of the single stage amplifier shown in fig. 3.28. An npn transistor that has $h_{ie} = 500$ Ω, h_{re} negligible, $h_{fe} = 150$, and $h_{oe} = 50$ μS is used. Neglect all coupling and stray capacitances.

(*b*) Comment on the various methods available for biasing the base and discuss their merits.

(Sheffield University: First year)

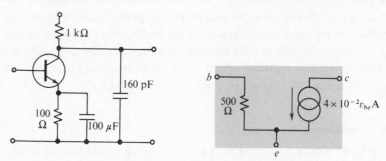

Fig. 3.29. Circuit for problem 3.

3. Fig. 3.29 shows the circuit of a single transistor amplifying stage and a circuit model of the transistor. Calculate the voltage gain at zero frequency and the approximate gain at 10 kHz. Calculate also the two frequencies at which the gain falls to 0.707 times the value which it has at 10 kHz.

(Cambridge University: First year)

4. Repeat problem 1(*b*) to show that the same results are given using the T-network model for the BC107 transistor. The parameters are;

$$\beta = 250, \quad r_d = 3.3 \text{ k}\Omega, \quad r_e = 1 \ \Omega, \quad r_b = 750 \ \Omega.$$

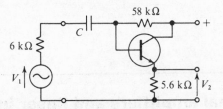

Fig. 3.30. Circuit for problem 5.

5. Fig. 3.30 shows the circuit of a common-collector transistor amplifier. The transistor used is a silicon type for which:

$h_{ie} = 2000\ \Omega$, h_{fe} ranges from 49 to 9, h_{oe} and h_{re} are negligible.

Calculate the maximum and minimum voltage gains V_2/V_1 and the output impedance of the circuit. The reactance of C is negligible at all relevant frequencies.

(Cambridge University: Second year)

6. A signal source has a maximum amplitude of ± 100 mV and frequency components which lie within the range 0·1 to 1000 Hz. A transistor amplifier is required which will magnify this signal voltage linearly so that a peak current of 10 mA can be supplied to a resistive load of 600 Ω.

Draw a circuit for a suitable amplifier and calculate the component values.

(Newcastle University: Third year)

4

Operational amplifiers and linear integrated circuits

4.1 Introduction

'Linear integrated circuits' is the name given to ready-built, miniature amplifier circuits which can be used directly in a number of applications often with the addition of only a few passive external components. This contrasts with the approach described in the preceding chapters where active devices such as field-effect transistors and bipolar transistors were used together with a number of passive circuit components to form amplifier circuits. This latter approach has come to be known as discrete circuit technology, as opposed to the integrated circuit technology which we describe in this chapter. The technical advantages of integrated circuits, coupled with the convenience with which they may be used, are of such value that their use has become extremely widespread. Circuits using them are much smaller than those made with discrete components and today they are cheaper too. With integrated circuits, the basic building blocks of electronic circuits which used to be collections of separate devices such as transistors, diodes, resistors, etc. have been miniaturised into ready-built circuits which can be easily linked together to form the desired system.

In most applications of integrated circuits a high gain amplifier forms the *internal* amplifier to which passive circuit elements are connected to provide an *overall* amplifier circuit called an 'operational amplifier'. (Note that the terminology is sometimes confusing in that the internal amplifier is itself referred to as the operational amplifier.) We shall first describe the realisation, characteristics, and uses of these operational amplifiers before turning in the latter sections of this chapter to a discussion of the design and fabrication of integrated circuits themselves.

4.2 A review of the properties of real and ideal amplifiers

One of the most important properties required from the basic internal amplifier to be used as the heart of an operational amplifier is that it must

114

have a very high gain. This specification is often extended to the assumption that the gain is high enough to be considered infinite for purposes of analysis and design. Further simplification of the analysis results from assumptions which can be made regarding other parameters of the amplifier. Ideally one assumes that the internal amplifier of an operational amplifier has infinitely high *input impedance* and a zero *output impedance*. Often it might also be required that these properties be maintained over a particular frequency range or at least be true, to a close approximation, at a specified operating frequency. It might also be necessary to specify that the performance of the amplifier be immune to changes in environmental conditions such as temperature. Real linear integrated circuit amplifiers cannot, of course, meet the specifications of our idealised amplifier but we look for the following properties in any amplifier we might select: high gain, a direct-current connection, a wide bandwidth, a high input impedance, a low output impedance, a high degree of stability against temperature and other environmental changes and a minimum of adjustment to set the correct DC conditions. To illustrate the degree to which these requirements are met by a typical device the main performance specifications of the μA709, a general purpose linear integrated circuit amplifier, are as follows:

> Voltage gain: 45000 times.
>
> Output resistance: 150 Ω.
>
> Input resistance: 250 kΩ.
>
> Operating temperature range: 0–70 °C.
>
> Offset requirements: input offset voltage 2.0 mV,
> input offset current 100 nA.

4.3 The terminology of integrated circuit differential amplifiers

The essential circuit connections to an integrated circuit, linear amplifier are (*a*) the power supply (usually both positive and negative voltage supplies with respect to ground or common rail are required), (*b*) the signal input and output terminals, (*c*) in some cases additional points are provided to which external circuit elements must be added as specified by the manufacturers, and (*d*) additional points to which are connected resistances or potentiometers to correct any offset DC levels associated with the input or output of the amplifier. (These offset levels are defined in §4.10.)

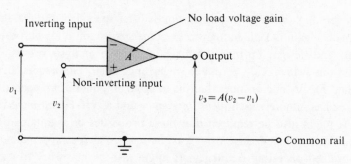

Fig. 4.1. Nomenclature.

There are usually two signal input terminals known as the 'inverting' and 'non-inverting' inputs and marked $(-)$ and $(+)$ respectively on the amplifier circuit symbol as shown in fig. 4.1. There is usually a single output, v_3 which is related to the inputs v_1, and v_2 by

$$v_3 = A(v_2 - v_1), \tag{4.1}$$

where A is the no-load voltage gain. The symbols indicate that a *negative* going signal applied to the $(-)$ terminal will produce a *positive* going output signal while at the $(+)$ terminal a *positive* going signal will again

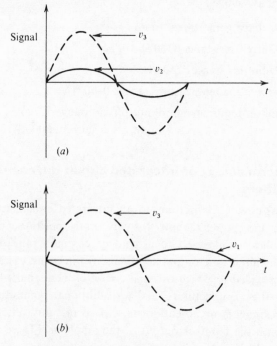

Fig. 4.2. Non-inverted and inverted signals.

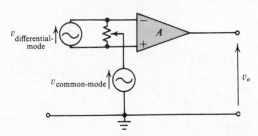

Fig. 4.3. Differential- and common-mode signals.

produce a *positive* going output. For sinusoidal input signals there is a 180° phase shift between the (−) input and the output and ideally a zero phase shift between the (+) terminal and the output, fig. 4.2. Signals may either be applied between the input points and the common rail or between the two inputs. The output is taken between the output terminal and the common rail.

An important property of the amplifier is that when identical signals are applied to *both* the (−) and the (+) terminals the output should be zero. This type of signal input to the amplifier is known as a *common-mode* signal input. When the signal is applied *between* the two input terminals it is known as a *differential-mode signal* and the amplifier produces a large output signal. This aspect is illustrated further in §4.9 where differential amplifiers are described. Fig. 4.3 illustrates the application of common-mode and differential-mode signals to the amplifier.

The conclusions so far are summarised as follows: the output from the amplifier is generally, $v_3 = A(v_2 - v_1)$. Two particular cases are; when $v_1 = v_2$, then $v_3 = 0$; when $v_2 = -v_1$ then $v_3 = -2A(v_1)$.

4.4   The inverting operational amplifier

4.4.1   Gain

We have stated that the basic high gain, integrated circuit amplifier may be connected to provide an amplified output which is inverted with respect to the input. However the large gain of the amplifier is not often used in the sense that, with no signal at the (+) input, a signal v_1 at the (−) input results in an output $-Av_1$. Instead the gain between input and output is almost invariably made substantially independent of the internal amplifier's gain and determined by accurate and stable passive circuit elements. The circuit usually employed is shown in fig. 4.4. The input signal is applied to the (−) terminal via a resistance R_1 often called the input resistor and the (+) terminal is connected to the common rail, therefore no signal voltage

117

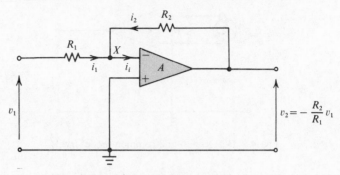

Fig. 4.4. The inverting amplifier.

appears at that input. The amplifier output is fed back to the $(-)$ terminal via R_2, called the feedback resistor. To calculate the overall gain, G, for the amplifier we assume first some rather unrealistic conditions, that the internal gain A is very large, the input resistance of the internal amplifier is infinite and the output resistance is zero. Later we examine what errors are made in using these assumptions.

If the amplifier gain A is very large, then, for a given output signal, the signal at the $(-)$ terminal is very small and the terminal is said to be at '*virtual earth*'. This concept is very useful in analysing the performance of operational amplifiers.

Returning to Fig. 4.4 we have at the node X, $i_1 = -i_2$ since the input impedance of the amplifier is infinite, so current into the amplifier, $i_i = 0$. The terminal X is at a virtual earth and therefore the equation for the currents becomes,

$$\frac{v_1}{R_1} = -\frac{v_2}{R_2}$$

and the overall gain $G = v_2/v_1 = -R_2/R_1.$ (4.2)

With the assumptions made here the overall gain of the operational amplifier has been shown to be independent of the internal amplifier gain A and determined entirely by the ratio of the external resistors R_2 and R_1 which are added to make up the overall amplifier.

Two of the assumptions made in this analysis, namely that gain, A, and input resistance, R_i, are very large, can be more critically examined by deriving an expression for the overall gain when R_i and A are finite. Fig. 4.5(a) shows the amplifier including R_i. The currents at the inverting input, X, to the internal amplifier are again summed. When the voltage at X is v',

$$\frac{v_1 - v'}{R_1} + \frac{v_2 - v'}{R_2} = \frac{v'}{R_i}.$$ (4.3)

118

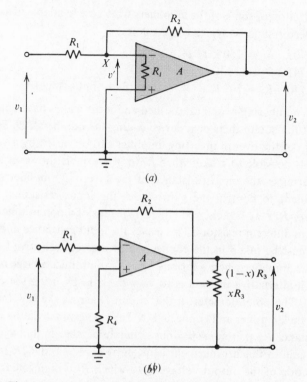

Fig. 4.5. (*a*) The inverting amplifier including input resistance.
(*b*) Gain adjustment.

But
$$v' = -v_2/A,$$
whence we have

$$\frac{v_2}{v_1} = -\frac{R_2}{R_1}\left[\frac{1}{1+\dfrac{1}{A}\left(1+\dfrac{R_2}{R_1}+\dfrac{R_2}{R_i}\right)}\right]. \tag{4.4}$$

Note that the term in the [ ] brackets will be approximately unity if the gain A is very large. Equation (4.4) then becomes (4.2).

How large A and R_i have to be depends on the numerical values used in any particular case under consideration and the accuracy of the result required. Some typical values might, for example, be $R_1 = 0.1$ MΩ, $R_2 = 1$ MΩ, $R_i = 350$ kΩ, $A = 10^3$.

Then $v_2/v_1 = -R_2/R_1 = 10$ approximately from (4.2), or substituting in (4.4) gives a gain of 9.87, and an error of about 1.3 per cent is made in using the approximate relation.

In amplifiers where the gain has to be very precisely set by the ratio R_2/R_1 both A and R_i can be made considerably larger than the values used

119

here. A selection of some of the amplifiers which are available in ascending order of performance is typically as follows:

μA709 $R_i = 250\text{ k}\Omega$ $A = 45\,000$

μA741 $R_i = 2\text{ M}\Omega$ $A = 2 \times 10^5$

μA740 $R_i = 10^{12}\,\Omega$ $A = 10^6$ (f.e.t. input stage).

This series numbering is well established and most makes have equivalents to these or include in their own device coding the numbers 709, 741, etc.

The gain of the overall inverting amplifier can be varied by a variety of means. It is possible to make either R_1 or R_2 variable, however R_1 effectively determines the input resistance of the inverting amplifier and thus its adjustment to control the gain causes the input resistance to vary. Variation of R_2 may be achieved either by means of a potentiometer or by switching in different resistors. This leaves the input resistance undisturbed but care must be taken in the connection to the very sensitive $(-)$ input point of the amplifier where spurious signals can produce a large unwanted output. An alternative method is to vary the gain by using the circuit of fig. 4.5(*b*). The potentiometer in this circuit is across the low impedance output of the amplifier and connected to the $(-)$ terminal via the relatively large resistance R_2, therefore spurious signals affect the amplifier very much less. The gain in this arrangement is given by $G = -1/x(R_2/R_1)$, where x is the fraction of the output voltage v_2 to which R_2 is connected.

A resistance $R_4 = R_1 \| R_2$ (R_1 in parallel with R_2) may be inserted from the non-inverting input to the common rail (see fig. 4.5(*b*)). Its purpose is to reduce the drift in amplifier gain because of temperature changes. The value of R_4 is determined by separate considerations to those involved in gain calculations and it does not affect the result for gain. This aspect is considered again in greater detail in §4.10 of this chapter which is concerned with offset balancing.

The considerations so far have taken into account the effect on gain of R_i and A being finite but the output impedance, R_o, has been assumed to be zero. It is straightforward to include the effects of R_o and also any additional load resistance placed on the amplifier. In fig. 4.6 the amplifier is shown again with a resistance R_o included in the output. The equations are now:

At node X,
$$\frac{v_1 - v'}{R_1} + \frac{v_2 - v'}{R_2} = \frac{v'}{R_i}. \tag{4.5}$$

At node Y,
$$\frac{v_o - v_2}{R_o} = \frac{v_2 - v'}{R_2}, \tag{4.6}$$

where
$$v_o = -Av'. \tag{4.7}$$

120

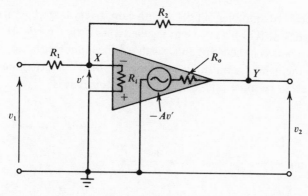

Fig. 4.6. Inverting amplifier including input and output resistance.

Eliminating v_o from (4.6) and collecting terms in (4.5) and (4.6)

$$\frac{v_1}{R_1} + \frac{v_2}{R_2} = v'\left(\frac{1}{R_i} + \frac{1}{R_2} + \frac{1}{R_1}\right), \tag{4.8}$$

$$\frac{v_2}{R_2} + \frac{v_2}{R_o} = v'\left(\frac{1}{R_2} - \frac{A}{R_o}\right). \tag{4.9}$$

Eliminating v' from (4.8) and (4.9) we have

the gain $G = \dfrac{v_2}{v_1} = \dfrac{R_i R_o - A R_2 R_i}{A R_1 R_i + R_o R_1 + R_o R_i + R_1 R_2 + R_1 R_i + R_i R_2}.$ (4.10)

This exact expression reduces to the form taken by (4.2) if A is very large, because then all the terms in the equation except those containing A, may be neglected and so

$$\frac{v_2}{v_1} \approx -\frac{R_2}{R_1}$$

which is independent of the internal amplifier's properties.

Typical figures for the output impedance, R_o, are: $R_o = 75\,\Omega$ for the μA740 and μA741 amplifiers and $R_o = 150\,\Omega$ for μA709. These values are so much less than the usual values of R_2 and R_1 that the effect of R_o on the amplifier gain may be neglected without appreciable error in nearly all cases.

4.4.2 Input and output impedances of the inverting amplifier

So far we have considered the gain of the operational amplifier, first by considering it to be based on an idealised internal amplifier, and then taking into account more and more the true performance of practical amplifiers.

The idealised amplifier of fig. 4.4, with gain A very large, has an extremely small signal at X which we have called the *virtual earth*. Hence R_1 is connected between the input and 'earth' so the input resistance, $v_1/i_1 = R_1$.

When the gain A, and input resistance R_i of the internal amplifier are considered to be finite (fig. 4.5(a)) the input current is

$$i_1 = \frac{v_1 - v'}{R_1},$$

and putting $v' = -\dfrac{v_2}{A}$, $\qquad i_1 = \dfrac{v_1}{R_1} + \dfrac{v_2}{AR_1}.$ $\qquad\qquad$ (4.11)

From (4.4), $\qquad v_2 = -v_1 \dfrac{R_2}{R_1} \left[\dfrac{1}{1 + \dfrac{1}{A}\left(1 + \dfrac{R_2}{R_1} + \dfrac{R_2}{R_i}\right)} \right],$

$$i_1 = v_1 \left[\frac{1}{R_1} - \frac{R_2}{R_1^2}\left(\frac{1}{A + 1 + \dfrac{R_2}{R_1} + \dfrac{R_2}{R_i}} \right) \right] \qquad\qquad (4.12)$$

whence $\qquad \dfrac{v_1}{i_1} = R_{in} = R_1 \left[1 + \dfrac{R_2 R_i}{R_1(R_i + AR_i + R_2)} \right].$ $\qquad\qquad$ (4.13)

In (4.13), A appears in the denominator of the second term in the brackets and will make that term very small, much smaller than unity, and the input resistance will be very close to R_1 in practice.

The output impedance is derived for the circuit of fig. 4.7 which assumes that R_i is very large. The currents at the nodes X and Y are written as:

$$\frac{v_1 - v'}{R_1} = \frac{v' - v_2}{R_2} \qquad\qquad (4.14)$$

and $\qquad \dfrac{-Av' - v_2}{R_o} + \dfrac{v' - v_2}{R_2} = i.$ $\qquad\qquad$ (4.15)

From (4.14) $\qquad v' = \dfrac{R_1 R_2}{R_1 + R_2}\left(\dfrac{v_1}{R_1} + \dfrac{v_2}{R_2} \right).$

Eliminating v' from (4.15) we have

$$v_2 = -\frac{v_1(AR_2 - R_o)}{R_o + R_1 + R_2 + AR_1} - i\frac{(R_o)(R_1 + R_2)}{R_o + R_1 + AR_1 + R_2}. \qquad\qquad (4.16)$$

Now compare (4.16) with

$$v_2 = v_1 \text{ (no-load voltage gain)} - i(\text{amplifier output resistance}).$$

122

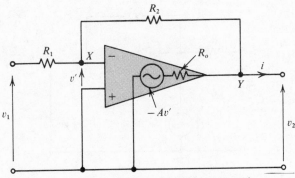

Fig. 4.7. Inverting amplifier including output resistance.

The term for no load voltage gain becomes $-R_2/R_1$ when A is large. Also the output resistance R_o' of the overall amplifier is

$$\frac{R_o(R_1+R_2)}{R_o+R_1(1+A)+R_2}$$

which is much less than R_o when A is large. For example, the amplifier, μA741, has $R_o = 75\,\Omega$ and $A = 2 \times 10^5$, so with $R_1 = 1$ kΩ and $R_2 = 10$ kΩ, we have $R_o' = 0.004\,\Omega$. Thus the output resistance of an inverting amplifier may be reduced to a very low value indeed.

4.5 The non-inverting operational amplifier

The non-inverting amplifier connection is shown in fig. 4.8. Here the signal is applied to the ($+$) terminal but the circuit connections using R_2 and R_1 are still made to the ($-$) terminal. One end of R_1 is joined to earth. This arrangement is necessary for reasons which will be clear when the chapters on negative and positive feedback have been studied.

4.5.1 Gain and output impedance

The case where $R_i \to \infty$ and $R_o \to$ zero is considered first. The output voltage $v_2 = A(v_1 - v')$ where v' is the voltage at the inverting input.

$$v' = \frac{R_1}{R_1+R_2}\,v_2,$$

$$\frac{v_2}{v_1} = \frac{A}{1+AR_1/(R_1+R_2)}. \tag{4.17}$$

When A is very large

$$\frac{v_2}{v_1} \approx 1+\frac{R_2}{R_1}. \tag{4.18}$$

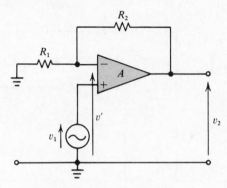

Fig. 4.8. Non-inverting amplifier.

Note that there is no negative sign here, hence the output and input signals are in phase. Again we have v_2/v_1 independent of the internal amplifier's parameters providing it is idealised to have infinite gain and input impedance and zero output impedance.

The non-inverting amplifier is now considered with a finite output resistance R_o, input resistance R_i and gain A, as shown in fig. 4.9. The amplifier is assumed to be supplied by a low impedance signal generator v_1 and feeds a current i_2 to a load.

The equations at nodes X and Y are:

$$\frac{v_2 - v'}{R_2} = \frac{v' - v_1}{R_i} + \frac{v'}{R_1}, \tag{4.19}$$

$$\frac{A(v_1 - v') - v_2}{R_o} = \frac{v_2 - v'}{R_2} + i_2. \tag{4.20}$$

Eliminating v' and simplifying we find that,

$$v_2 = v_1 \left(\frac{R_o R_1 + A(R_i R_2 + R_1 R_i)}{R_i(R_1 + R_2 + A R_1) + R_1 R_2 + R_o R_i + R_1 R_o} \right)$$

$$- i_2 \left(\frac{R_o(R_i R_2 + R_1 R_2 + R_1 R_i)}{(1 + R_o/R_2)(R_i R_2 + R_1 R_2 + R_1 R_i) + A R_i R_1 - R_o R_1 R_i/R_2} \right). \tag{4.21}$$

We can compare this with the expression for the output voltage v_2 of an amplifier,

$$v_2 = v_1 \text{ (no-load voltage gain)} - i_2 \text{ (output resistance)},$$

where v_1 is the input voltage and i_2 is the output current.

Examining the term that represents the no-load voltage gain in (4.21),

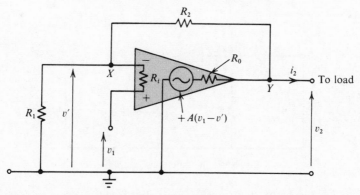

Fig. 4.9. Non-inverting amplifier including input and output resistance.

when A is large all the terms in the expression are relatively small except those containing A. Therefore,

$$\text{No-load voltage gain} \approx \frac{AR_i R_2 + AR_1 R_i}{AR_i R_1} \approx 1 + \frac{R_2}{R_1}.$$

Examining the term that represents output resistance and again assuming A is large we have:

$$\text{Output resistance} \approx \frac{R_o(R_i R_2 + R_1 R_2 + R_1 R_i)}{AR_i R_1}$$

$$\approx \frac{R_o}{A}\left(1 + \frac{R_2}{R_1}\right) \text{ when } R_i \text{ is large.} \quad (4.22)$$

Thus the output resistance is very low since $(1 + R_2/R_1) \ll A$.

4.5.2 Input impedance

The input resistance R_{in} of the non-inverting amplifier can be made very much larger than R_i, the resistance between the $(-)$ and $(+)$ terminals.

We have

$$R_{in} = v_1/i_1,$$

where the input current, $\quad i_1 = (v_1 - v')/R_i,$

therefore $\qquad\qquad R_{in} = \dfrac{R_i}{(1 - v'/v_.)}.$

On no load we have $i_2 = 0$ and (4.19) and (4.20) may be used to eliminate v'/v_1 to give,

$$R_{in} = R_i + R_1 \left(\frac{R_2 + R_o + AR_i}{R_1 + R_2 + R_o}\right). \quad (4.23)$$

When R_o is small and A is large

$$R_{in} \approx R_i + \frac{AR_i}{1 + R_2/R_1} \approx \frac{AR_i}{1 + R_2/R_1}. \quad (4.24)$$

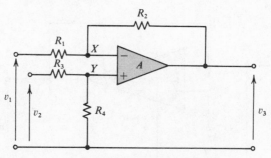

Fig. 4.10. The differential amplifier.

Since $A \gg 1 + R_2/R_1$ the input resistance is much greater than for the inverting amplifier, e.g. the μA741 with $R_2/R_1 = 100$ gives $R_{in} = 4 \times 10^9 \, \Omega$. In practice one is not always able to realise such a high input resistance because it is shunted by another resistance which is called the common-mode input resistance. This latter resistance is found between the $(-)$ and $(+)$ inputs and common rail or earth. Its value is sometimes specified by manufacturers and can vary between 10^7 and 10^{12} ohms.

Shunting the input resistance is a capacitance which arises from the stray, or sometimes, deliberately added, capacity between the output of the amplifier and the $(-)$ or inverting terminal. This capacity is multiplied by the gain and the input impedance is reduced at high frequencies. Further discussion of these topics is beyond the scope of this introductory text.

4.6 The differential operational amplifier

The inverting and non-inverting inputs of the internal amplifier have been used separately in the two configurations described so far in §§4.4 and 4.5. It is possible to use the amplifier in a form where signals are applied to both inputs. This is known as the differential or difference amplifier mode. (The design and operation of differential amplifiers using discrete components are described in §4.9.) The basic property of this mode of operation is that it amplifies the *difference* between the two input signals. In its idealised form it gives zero output for identical signals applied at the two inputs. The circuit of the basic differential operational amplifier is shown in fig. 4.10. The two input signals are v_1 and v_2. The output voltage, v_3, may again be determined by writing the nodal equations for the points marked X and Y at which the potentials v_1' and v_2' are assumed.

At nodes X and Y

$$\frac{v_1 - v_1'}{R_1} = \frac{v_1' - v_3}{R_2} \quad \text{and} \quad \frac{v_2 - v_2'}{R_3} = \frac{v_2'}{R_4}$$

126

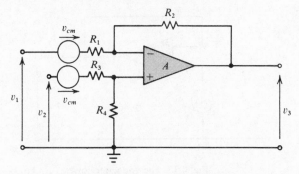

Fig. 4.11. Common-mode input to a differential amplifier.

and $$A(v_2' - v_1') = v_3.$$

Eliminating v_1' and v_2' and assuming A is large gives

$$v_3 = \frac{R_4}{R_3}\left(\frac{1 + R_2/R_1}{1 + R_4/R_3}\right)v_2 - \left(\frac{R_2}{R_1}\right)v_1. \tag{4.25}$$

The resistance ratios R_2/R_1 and R_4/R_3 can be made equal and the output then simplifies further to

$$v_3 = \frac{R_2}{R_1}(v_2 - v_1). \tag{4.26}$$

Therefore v_3 is an amplified version of the *difference* of the two input signals v_2 and v_1.

In differential amplifiers we usually wish to eliminate all traces from the output of signals which are identical at the two input terminals. This signal is known as the common-mode signal. Physically it might be a spurious pick-up signal, a temperature induced drift or a power supply fluctuation at the two inputs or any other undesired signal input. Fig. 4.11 shows the amplifier with a common-mode signal, v_{cm} in addition to the input signals v_1 and v_2. We can use the superposition theorem to calculate the additional output due to the common-mode signal with $v_1 = v_2 = 0$. Then we have

$$\frac{v_{cm} - v_1'}{R_1} = \frac{v_1' - v_3}{R_2}$$

and $$\frac{v_{cm} - v_2'}{R_3} = \frac{v_2'}{R_4} \quad \text{and} \quad A(v_2' - v_1') = v_3.$$

Eliminating v_2' and v_1' we have, assuming A is large,

$$v_3 = v_{cm}\frac{R_4 R_2 + R_4 R_1 - R_2 R_3 - R_2 R_4}{R_1(R_3 + R_4)}. \tag{4.27}$$

The output from the signals v_1 and v_2 in the absence of v_{cm} was given

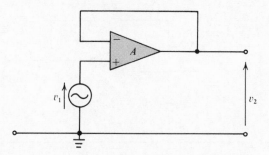

Fig. 4.12. The voltage follower.

by (4.25). Therefore the total output when v_1, v_2 and v_{cm} are all present, as is usually the case, is

$$v_3(\text{total}) = \frac{R_4(1 + R_2/R_1)}{R_3(1 + R_4/R_3)} v_2 - \left(\frac{R_2}{R_1}\right) v_1 + \left(\frac{R_4 R_2 + R_4 R_1 - R_2 R_3 - R_2 R_4}{R_1(R_3 + R_4)}\right) v_{cm}.$$

(4.28)

If the ratio R_2/R_1 is very close to R_4/R_3 then the coefficient of v_{cm} in (4.28) will be very small while the first part of the expression will yield approximately the difference signal as given in (4.26).

It follows that when the resistance ratios are chosen carefully then the amplifier has the property of producing zero or a very small output for identical signals at both inputs and a relatively large output for the difference input signals.

4.7 Some applications of operational amplifiers

4.7.1 The voltage follower

A circuit which has the very useful property of providing the highest possible input impedance is known as the voltage follower. Its voltage gain is very close to unity and its main function is to act as isolation between a signal generator and a load resistance, hence it is also known as the unity-gain buffer amplifier. The circuit is shown in fig. 4.12 and its analysis may either be treated as a special case of the non-inverting amplifier, or alternatively we write the equation for output voltage as $A \times$ (difference of inputs)

$$v_2 = A(-v_2 + v_1).$$

Therefore

$$v_2/v_1 = \frac{A}{1+A}.$$

(4.29)

When $A \gg 1$ this is very close to unity, and v_2 equals v_1 to a very close approximation.

128

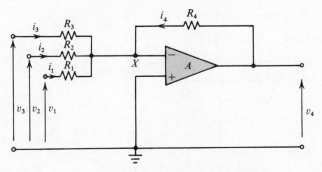

Fig. 4.13. The adding/scaling amplifier.

When the internal amplifier has an input resistance R_i between the $(-)$ and $(+)$ inputs, the input current is $(v_1 - v_2)/R_i$, so,

$$R_i(\text{overall}) = \frac{v_1}{(v_1 - v_2)/R_i} = \frac{v_1 R_i}{v_1 - A v_1/(1+A)} = (1+A)R_i \quad (4.30)$$

which can be very large when both A and R_i are large as is usually the case.

If the internal amplifier has an output resistance R_o then the output resistance of the overall amplifier, the voltage follower, is $R_o/(1+A)$.

Thus the properties of this circuit are a voltage gain very close to unity, a very high input resistance and a low output resistance. The properties compare with those of the source follower and emitter follower circuits described in chapters 2 and 3 respectively.

4.7.2 The adder amplifier

An amplifier with properties approaching the idealised criteria, viz.

$$\text{Gain } A \to \infty, \quad R_i \to \infty \quad \text{and} \quad R_o \to 0,$$

may be used to add a number of signal inputs accurately. The circuit is shown in fig. 4.13. The node X has the currents i_1, i_2, and i_3 flowing in through the resistors R_1, R_2 and R_3 and a current i_4 flowing through R_4 from the output terminal. There could be more inputs and more terms in the equations summing the currents. As X is a virtual earth, we have,

$$\frac{v_4}{R_4} + \left(\frac{v_1}{R_1} + \frac{v_2}{R_2} + \frac{v_3}{R_3}\right) = 0$$

or

$$v_4 = -R_4\left(\frac{v_1}{R_1} + \frac{v_2}{R_2} + \frac{v_3}{R_3}\right). \quad (4.31)$$

Thus the voltages v_1, v_2, etc. are scaled individually and summed up. Alternatively, it is possible to make $R_1 = R_2 = R_3 = R_4$ and we have

$$v_4 = -(v_1 + v_2 + v_3). \quad (4.32)$$

129

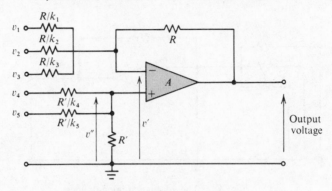

Fig. 4.14. Adding/scaling–subtracting/scaling amplifier.

It is interesting to determine the overall gain when the amplifier gain A and input resistance R_i are finite. Then at the node X (fig. 4.13) whose voltage is $-v_4/A$, the currents are,

$$\frac{v_1-(-v_4/A)}{R_1}+\frac{v_2-(-v_4/A)}{R_2}+\frac{v_3-(-v_4/A)}{R_3}+\frac{v_4-(-v_4/A)}{R_4}=\frac{-v_4/A}{R_i}$$

or $$v_4\left[\frac{1}{R_4}+\frac{1}{A}\left(\frac{1}{R_1}+\frac{1}{R_2}+\frac{1}{R_3}+\frac{1}{R_4}+\frac{1}{R_i}\right)\right]=-\left(\frac{v_1}{R_1}+\frac{v_2}{R_2}+\frac{v_3}{R_3}\right). \quad (4.33)$$

Compared with the idealised amplifier the term,

$$\frac{1}{A}\left(\frac{1}{R_1}+\frac{1}{R_2}+\frac{1}{R_3}+\frac{1}{R_4}+\frac{1}{R_i}\right)$$

introduces an error in the scaling and adding operation. The extent of the error depends on the specification of the amplifier chosen. For

$$R_1 = R_2 = R_3 = R_4 = 1 \text{ M}\Omega,$$

and using a μA709 amplifier with $R_i = 250$ kΩ and $A = 25000$

$$v_4(10^{-6}+0.32\times10^{-3}) = -(v_1+v_2+v_3)10^{-6},$$

$$v_4 = -\frac{1}{1.00032}(v_1+v_2+v_3).$$

There is an error of just over 3 parts in 10^4.

By making use of both the $(+)$ and the $(-)$ inputs it is possible to have an adder/subtractor amplifier as shown in fig. 4.14. A similar analysis to that for the adder shows that if

$$k_1+k_2+k_3 = k_4+k_5,$$

output voltage $= k_4v_4+k_5v_5-(k_1v_1+k_2v_2+k_3v_3).$

130

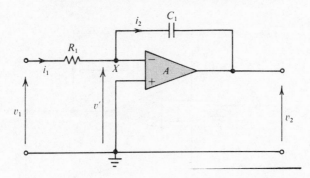

Fig. 4.15. Integrating amplifier.

So the inputs v_4 and v_5 are scaled and summed and the inputs v_1, v_2 and v_3 are scaled and subtracted from them.

4.7.3 Integrating and differentiating amplifiers

An operational amplifier with a capacitor in place of the resistance which usually links output to input acts as an integrating circuit. (The circuit is shown in fig. 4.15.) The relation between input and output voltages is again determined by summing the currents at the node X where a voltage v' is assumed

$$\frac{v_1 - v'}{R_1} = i_1 = i_2.$$

Since i_2 flows through the capacitor C_1, we can write

(capacitor voltage) × its capacitance = charge,

or

$$v_2 - v' = -\frac{1}{C_1} \int_0^t i_1 \, dt.$$

But

$$v' = -v_2/A.$$

When A is very large $v' \to 0$

and

$$v_2 = -\frac{1}{R_1 C_1} \int_0^t v_1 \, dt. \tag{4.34}$$

The output is the integral of the input voltage v_1. By modifying the circuit as shown in fig. 4.16 it is possible to include an initial condition together with the integral. The output from this circuit is

$$v_2 = -\frac{1}{R_1 C_1} \int_0^t v_1 \, dt - \left(\frac{R_B}{R_A}\right) V_B. \tag{4.35}$$

The differentiating operational amplifier may be constructed by merely interchanging the positions of the capacitor and resistors of the integrator to the positions shown in fig. 4.17.

131

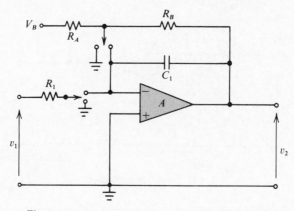

Fig. 4.16. Integrating amplifier with initial conditions.

The output is then given by

$$v_2 = -R_1 C_1 \frac{\mathrm{d}v_1}{\mathrm{d}t}.$$

This circuit is however rarely used because it suffers from the disadvantage that it amplifies noise and spurious voltage spikes at the input. A capacitor across R_1 as indicated in fig. 4.17 can be used to minimise this effect at the expense of exact differentiation.

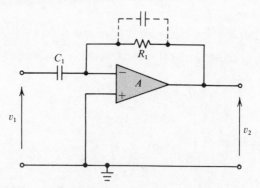

Fig 4.17. Differentiating amplifier.

4.7.4 Simulation of an inductance

Inductance is avoided in integrated circuit design because of technological difficulties. In discrete circuit design also, inductance usually means coils which are relatively large when compared with the other components. An interesting application of an operational amplifier is to simulate an inductance by using an R–C network in conjunction with the amplifier.

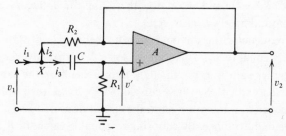

Fig. 4.18. Simulation of an inductance with an operational amplifier.

The circuit of fig. 4.18 can be analysed to show that its input impedance $Z_i = v_1/i_1$ is of the form $R+j\omega L$ which means that its inductance is L and its Q or 'quality factor' is $\omega L/R$.

The gain of the operational amplifier connected with its inverting input as shown in fig. 4.18 has been determined in §4.7.1. It is a voltage follower circuit of voltage gain, very close to unity, very high input impedance and very low output impedance provided the open loop gain A is large.

We have therefore $v_2 \approx v' \approx v_1 R_1/(R_1 + 1/j\omega C)$.

Summing the currents at the node X

$$i_1 = i_2 + i_3,$$

and, since the amplifier draws a negligible current,

$$i_1 = \frac{v_1 - v_2}{R_2} + \frac{v_1}{R_1 + 1/j\omega C}.$$

Eliminating v_2 from the two equations we have

$$\frac{v_1}{i_1} = \frac{R_1 R_2 j\omega C + R_2}{1 + j\omega C R_2}. \tag{4.36}$$

Separating (4.36) into real and imaginary parts

$$\frac{v_1}{i_1} = Z_i = \frac{R_1 R_2{}^2 \omega^2 C^2 + R_2}{1 + \omega^2 C^2 R_2{}^2} + j\frac{\omega R_2 C}{1 + \omega^2 C^2 R_2{}^2}(R_1 - R_2).$$

For Z_i to appear an inductor $R_1 - R_2$ must be positive and in fact it would be usual to make $R_1 \gg R_2$. The Q of the circuit is

$$\frac{\omega C(R_1 - R_2)}{R_1 R_2 \omega^2 C^2 + 1}.$$

For the usual values of frequency of operation, the choice of components leads to $\omega^2 C^2 R_2{}^2 \ll 1$ and $R_1 \gg R_2$, therefore the inductance may be written as approximately $L \approx R_1 R_2 C$.

At high frequencies the circuit begins to behave as a pure resistance approximately equal to R_1. This follows since C is virtually a short circuit and the amplifier gain is approximately unity; therefore the same voltage appears at each end of R_2 which consequently draws very little current. At low frequencies on the other hand the circuit appears to be a resistance in series with an inductance. By choosing reasonable values of R_1, R_2 and C an inductance of a few henrys may be obtained at ~ 500 Hz.

4.7.5 A band-pass filter

A band-pass filter must pass all signals over a range of frequency without attenuation or perhaps even with amplification. All signals outside this range must be attenuated by as large a factor as possible.

It is possible to construct such a filter using frequency selective feedback across an amplifier. Unwanted signals are fed back to keep the overall gain low while the wanted signals are not fed back. A suitable frequency selective network is a twin-T network but it requires to have in it six carefully matched components to function well. It is also difficult and expensive to adjust the output signal from the circuit. Full details of this arrangement may be found in books on filters.

A simpler circuit which overcomes this difficulty and uses only five components in addition to the amplifier is shown in fig. 4.19. This circuit may be adjusted so as to obtain the desired centre frequency by the adjustment of one resistive element only. Moreover this adjustment leaves the gain undisturbed.

The gain v_2/v_1 for the circuit may be determined by summing the currents at the node X on the circuit of fig. 4.19:

$$i_1 + i_2 + i_3 = i_4$$

or
$$\frac{v_1 - v_3}{R_1} + \frac{-v_2/A - v_3}{1/j\omega C_0} + \frac{v_2 - v_3}{1/j\omega C_0} = \frac{v_3}{R_2}. \tag{4.37}$$

We assume that the input resistance of the operational amplifier is very high and a negligible current flows into it. Therefore

$$i_5 = \frac{v_2 + v_2/A}{R_3} = \frac{-v_2/A - v_3}{1/j\omega C_0}. \tag{4.38}$$

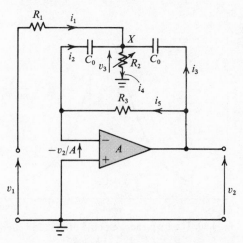

Fig. 4.19. A filter circuit.

From (4.38) we may express v_3 in terms of v_2

$$v_3 = -\left(\frac{v_2}{R_3} + \frac{v_2/A}{R_3} + (v_2/A)\,j\omega C_0\right) 1/j\omega C_0$$

$$= \frac{j v_2}{\omega C_0 R_3} - \frac{v_2}{A} \quad \text{since} \quad v_2/A \ll v_2.$$

Substituting this value of v_3 in (4.37) and simplifying we have:

$$\text{Gain } \frac{v_2}{v_1} = \frac{-1}{R_1\left\{\dfrac{2}{R_3} + \dfrac{1}{A}\left(\dfrac{1}{R_1} + \dfrac{1}{R_2}\right)\right\} + j R_1\left\{\dfrac{1/R_1 + 1/R_2}{\omega C_0 R_3} - \omega C_0\right\}}. \quad (4.39)$$

The gain will have a maximum when the imaginary term in the denominator is zero, which gives

$$\omega^2 = \frac{1/R_1 + 1/R_2}{C_0^{\,2} R_3} \quad \text{or} \quad f_0 = \frac{1}{2\pi C_0}\left(\frac{R_1 + R_2}{R_1 R_2 R_3}\right)^{\frac{1}{2}}. \quad (4.40)$$

The maximum gain corresponding to the centre frequency, f_0 is given by

$$\text{Gain}_{max} = \left|\frac{v_2}{v_1}\right|_{max} = \frac{-1}{\dfrac{2R_1}{R_3} + \dfrac{R_1}{A}\left(\dfrac{1}{R_1} + \dfrac{1}{R_2}\right)}. \quad (4.41)$$

The second term in the denominator is very small compared with $2R_1/R_3$, hence

$$\text{Maximum gain} \approx -R_3/2R_1. \quad (4.42)$$

135

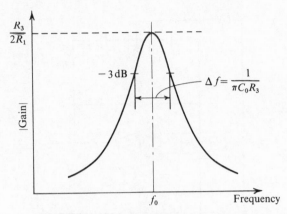

Fig. 4.20. The frequency response of the filter circuit

$$f_0 = \frac{1}{2\pi C_0} \sqrt{\frac{R_1 + R_2}{R_1 R_2 R_3}}.$$

Equating the real part to the imaginary part of (4.39) gives the half power bandwidth as

$$\Delta f = \frac{1}{\pi C_0 R_3} \tag{4.43}$$

and

$$Q = \frac{f_0}{\Delta f} = \tfrac{1}{2}(R_3/(R_1 \| R_2))^{\frac{1}{2}}. \tag{4.44}$$

The response of the circuit as a function of frequency is illustrated in fig. 4.20. The circuit has the advantages that its centre frequency may be adjusted conveniently by changing R_2 leaving the maximum gain value undisturbed since that is independent of R_2, (4.42). A range of centre frequencies from 10 Hz to 10 kHz is obtainable using this circuit.

4.7.6 Current-to-voltage converter

Certain electronic devices such as photoelectric cells are more effective and linear as current rather than voltage sources. This current signal can be converted with an operational amplifier to a voltage output. As shown in fig. 4.21 the current is injected into the amplifier node, X, which is connected to the inverting input and also to one end of the feedback resistor R_2. Since this is the 'virtual earth' point very little impedance is offered to the current source. The current flows into the resistor R_2 and the output voltage v_2 is shown to simplify to:

$$v_2 = i_o R_2.$$

The particular advantage of the circuit is its low input impedance so that

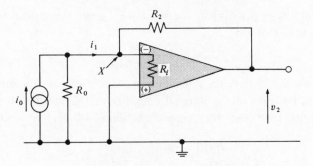

Fig. 4.21. Circuit of current to voltage converter.

this connection causes the least voltage to be developed at the input (condition for best loading of any current source). The input impedance of the op-amp can be found for an internal resistance R_i between the inverting and non-inverting terminals. It is obtained by summing the currents at the node X. If v' is the voltage at the $(-)$ input, then the input current i_1 is:

$$i_1 = \frac{v'}{R_i} + \frac{v' - v_2}{R_2}. \tag{4.45}$$

Since $v_2 = -Av'$, the input resistance is given by

$$R_{in} = \frac{v'}{i_1} = \frac{R_2 R_1}{R_2 + R_i(1 + A)}.$$

When A is large, the second term in the denominator predominates, so

$$R_{in} \approx R_2/(1 + A)$$

which is very small for large A. This is the ideal condition for best current coupling (chapter 1) so i_1 is very close to i_o. Putting this relation into equation (4.45) and putting $v' = v_2/(-A)$ gives the equation for v_2 quoted.

Thus the circuit gives a voltage output for a current input.

4.8 Frequency response of operational amplifiers

The gain of the integrated circuit amplifier has been treated so far, as being independent of frequency when in fact it will drop as the frequency rises. In chapters 2 and 3 we have already shown that with f.e.t. and bipolar transistor amplifiers, the gain beyond the upper half power frequency decreases by 20 dB/decade frequency rise. Many integrated circuit amplifiers perform in the same way, since the typical frequency response is given by:

$$A(f) = \frac{A_0}{1+j(f/f_0)},\qquad(4.46)$$

where the symbols have the same meaning as in previous chapters, namely that A_0 is the gain of the internal amplifier at DC and low frequencies and f_0 is the 'turn over' frequency. Equation (4.46) may be written as:

$$A(\text{dB}) = 20\log\frac{A_0}{[1+(f/f_0)^2]^{\frac{1}{2}}}$$

with $\qquad\qquad \phi = -\tan^{-1}f/f_0 \qquad\qquad(4.47)$

which may be plotted as shown in figs. 4.22 (*a*) and (*b*). The special features of these curves have been discussed in chapter 1. The voltage gain of an integrated circuit amplifier at low frequency is very large, 80 dB or more can be obtained when the circuit is loaded by a resistance equal to its input resistance. The response curves are shown as Amplifier I in fig. 4.22 (*a*) and (*b*) for gain and phase angle respectively. The (gain × bandwidth) product is shown in chapter 5 to be approximately constant and to some extent gain and frequency response may be traded against each other in operational amplifier circuits. For example if the resistances R_1 and R_2 are added to reduce the overall gain, as described in the previous sections of this chapter, by some specified factor then the bandwidth will be increased in the same proportion. This is illustrated by curves marked Amplifier II on figs. 4.22(*a*) and (*b*).

Many operational amplifiers have this standard type of frequency response. Some, however, have faster rates of fall in frequency. This has both advantages and disadvantages which we point out in greater detail in chapter 5. The main point to realise is that the maximum phase shift for the type of frequency response in fig. 4.22 is $-90°$ only. If a faster fall off in gain with frequency is employed then phase shifts approaching $-180°$ may be obtained. This can cause the amplifier to become unstable and allow oscillations to build up. The condition is discussed again in chapter 5.

The manufacturers of integrated circuit amplifiers may specify the behaviour of their product in terms of frequency in three different ways: (*a*) the 3 dB frequency may be specified, (*b*) the frequency at which the gain falls to 0 dB, i.e. unity voltage gain, may be specified or (*c*) the product of A_0 and the frequency at which the gain drops to 0 dB may be specified. Therefore care must be taken in interpreting manufacturer's data on frequency response.

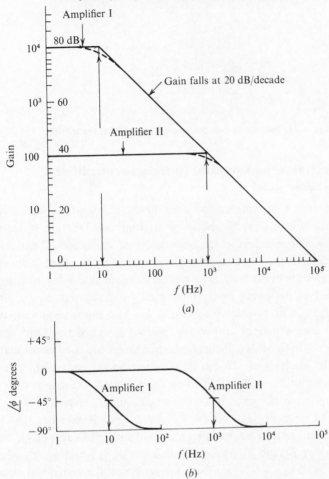

Fig. 4.22. Frequency response of the operational amplifier. (*a*) Gain as a function of frequency, (*b*) phase as a function of frequency.

Some integrated circuits require external frequency compensation components which must be added on to ensure that the correct frequency response is obtained. For example the circuit of the μA709 amplifier shown in fig. 4.23 requires a resistance–capacitance combination between pins 1 and 8 and a capacitance between pins 5 and 6 as compensating elements. The exact values of the components are specified by the manufacturers to achieve the required bandwidth (appendix B). It should also be noted that it is not only the gain which is a function of frequency. The input and output resistances as well as the input capacity also vary with frequency, typically in the manner shown in appendix B.

139

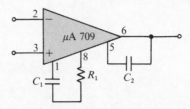

Fig. 4.23. Frequency compensation (external) for the μA 709.

4.9 The differential amplifier (difference amplifier, long-tailed pair)

So far we have not considered the amplifier circuit used within the triangular block which denotes the integrated circuit. Discrete transistor circuits, similar in principle to the ones described in chapters 2 and 3 are used. For example, the differential amplifier is used a great deal in discrete component circuits but it is of even greater consequence today because it is widely used in designing linear integrated circuit amplifiers. Its most striking feature is its symmetry in that it has two input terminals, two outputs and uses a pair of transistors, often a matched pair. The basic circuit is shown in fig. 4.24 (*a*) where matched f.e.t.s. are used as the active elements. The analysis of the circuit is based on the equivalent circuit diagram shown in fig. 4.24(*b*). The two input signals are assumed to be v_1 and v_2 applied to the gates of transistors T_1 and T_2 respectively.

We wish to write expressions for v_3/v_1 and v_4/v_2. Remember that the equivalent circuit of the f.e.t. includes a current source of value $g_m v_{gs}$, and, in this case, for T_1 and T_2, the generators are $g_m v_{gs1}$ and $g_m v_{gs2}$ where $v_{gs1} = v_1 - v_5$ and $v_{gs2} = v_2 - v_5$. These are shown on the equivalent circuit of fig. 4.24 (*b*).

For the nodal point S where the two f.e.t. sources are connected, we sum the currents to give,

$$g_m(v_1 - v_5) + \frac{(v_3 - v_5)}{r_d} + g_m(v_2 - v_5) + \frac{(v_4 - v_5)}{r_d} - \frac{v_5}{R_5} = 0. \qquad (4.48)$$

At the earth node marked E we have, on no load,

$$\frac{v_5}{R_5} + \frac{v_3}{R_3} + \frac{v_4}{R_4} = 0.$$

Summing the currents at the drain of T_1, marked D_1, we have

$$g_m(v_1 - v_5) + \frac{v_3 - v_5}{r_d} + \frac{v_3}{R_3} = 0, \qquad (4.49)$$

140

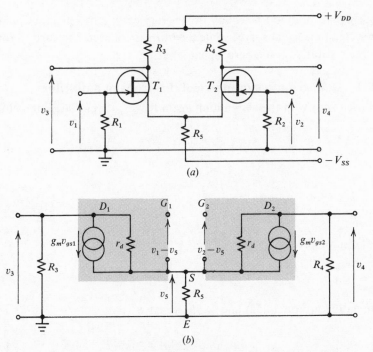

Fig. 4.24. (a) A f.e.t. differential amplifier. (b) Equivalent circuit.

and at the drain of T_2, marked D_2, we have

$$g_m(v_2 - v_5) + \frac{v_4 - v_5}{r_d} + \frac{v_4}{R_4} = 0. \qquad (4.50)$$

These equations may be solved for the two gain ratios v_3/v_1 and v_4/v_2. However, lengthy equations result if we do not make certain simplifications to the circuit of fig. 4.24. For instance, it is normal practice to design the circuit to be symmetrical. The two drain resistors are made equal, i.e. $R_3 = R_4$, and the common-source resistance R_5 is made as large as possible for reasons which will become apparent in this analysis.

Equation (4.48) may be simplified if R_5 is 1000 times R_4 or R_3, to:

$$\frac{v_3}{R_3} + \frac{v_4}{R_4} = 0 \quad \text{and when} \quad R_3 = R_4,$$

$$v_3 = -v_4.$$

Therefore the circuit produces output signals which are equal in magnitude but opposite in polarity or, in the case of sinusoidal signals, the two outputs are equal in magnitude but 180° out of phase with each other. The

141

input signals need not be equal. In fact very often only one input point is connected to a signal source while the other is connected to earth through a large resistor so it has zero signal input.

4.9.1 Gain from a symmetrical differential amplifier

We use (4.48) with the assumption again that R_5 is very large, therefore,

$$g_m(v_1+v_2)+\frac{(v_3+v_4)}{r_d}-v_5\left(2g_m+\frac{2}{r_d}\right) = 0 \tag{4.51}$$

but
$$v_3 = -v_4,$$

therefore
$$g_m(v_1+v_2) = v_5(2g_m+2/r_d). \tag{4.52}$$

In (4.49) we have

$$g_m v_1 + v_3(1/R_3+1/r_d)-v_5(g_m+1/r_d) = 0. \tag{4.53}$$

Eliminating v_5 from (4.52) and (4.53) we have

$$v_3 = -v_4 = \frac{-g_m(v_1-v_2)}{2(1/r_d+1/R_3)}. \tag{4.54}$$

The output, therefore, is proportional to the *difference* between the two input signals v_1 and v_2 which is one of the most significant properties of the circuit. Some interesting features may be observed from examination of (4.54).

(a) When $v_1 = v_2$, the outputs v_3 and v_4 are zero. $\qquad$ (4.55)

Although this result is trivial when v_2 and v_1 are the signal sources it is significant in the context of spurious components superimposed on the wanted signals v_1 and v_2. When these spurious input signals exist they are often identical while v_1 and v_2 are in fact different. The result of the spurious inputs would not appear at the amplifier output having been cancelled within the amplifier. This is another very important property of the amplifier.

The amplifier is particularly suitable as a DC amplifier since it compensates against drift caused by temperature changes. The temperature fluctuation would in general apply to both transistors equally leading to changes in drain current. This is equivalent to an equal change in v_1 and v_2 and cancels. By means of suitable precautions such as using a pair of matched

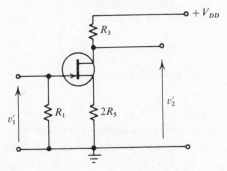

Fig. 4.25. One half of a symmetrical differential amplifier.

f.e.t.s one can reduce the difference voltage $(v_{gs1} - v_{gs2})$ for a temperature rise of 1 °C to as little as 5 μV.

(b) When $v_2 = 0$,

$$\text{output } v_3 = -v_4 = \frac{-g_m v_1}{2(1/r_d + 1/R_3)}. \qquad (4.56)$$

(c) When $v_1 = -v_2$

$$\text{output} = -\frac{g_m v_1}{(1/r_d + 1/R_3)}, \quad \text{i.e. double the result in (4.56).} \quad (4.57)$$

(d) If R_5 is made effectively a short circuit for AC by decoupling it, i.e. by putting a large enough capacitor in parallel with it, we have $v_5 = 0$.
Equation (4.49) then gives

$$g_m v_1 + \frac{v_3}{r_d} + \frac{v_3}{R_3} = 0,$$

therefore

$$v_3 = \frac{-g_m v_1}{1/r_d + 1/R_3}. \qquad (4.58)$$

Which shows on comparison with (4.54) that it is the presence of R_5 in the circuit, which gives the circuit its difference amplifier properties, and also reduces the gain by a factor of 2.

4.9.2 The common-mode rejection ratio (CMRR) of the differential amplifier

An important parameter defining the quality of a differential amplifier is its common-mode rejection ratio. This is defined as the ratio:

$$\frac{\text{voltage gain for difference signals, i.e. } v_1 = -v_2}{\text{voltage gain for common-mode signal, i.e. } v_1 = v_2}.$$

The difference or differential mode gain is obtained from (4.54) when

$$v_1 = -v_2 = v,$$

$$\text{gain (differential)} = \frac{-g_m}{(1/r_d + 1/R_3)}. \tag{4.59}$$

The common-mode gain cannot be determined from (4.54) in which we have assumed that R_5 is very large. It gives a result of zero output for equal inputs signals, which makes the common-mode gain zero. For any finite value of R_5, however, the common-mode gain is obtained by recognising the symmetry of the circuit and dividing it into two identical parts. This can be done because identical current components arise in R_5 from the two common-mode or equal input signals. Each half of the circuit now appears as shown in fig. 4.25 with R_3 in the drain lead but with $2R_5$ in the source lead since R_5 has to be divided into two parallel paths as it is shared between the two halves. For each half the gain is v_2'/v_1' and this by definition is the denominator of the common-mode rejection ratio. The gain is determined for the circuit of fig. 4.25 by the usual method of drawing the small-signal equivalent circuit and analysing for the ratio of the output and input voltages which gives

$$\frac{v_2'}{v_1'} = \frac{g_m R_3}{1 + 2g_m R_5 + 2R_5/r_d + R_3/r_d}. \tag{4.60}$$

Hence

$$\text{CMRR} = \frac{\text{gain (differential), equation (4.59)}}{\text{gain (common-mode), equation (4.60)}}$$

$$= \frac{1 + 2g_m R_5 + 2g_d R_5 + g_d R_3}{g_d R_3 + 1}, \tag{4.61}$$

where

$$g_d = 1/r_d.$$

As $g_d R_3 \ll 1$ in the denominator of (4.61), $2g_m R_5 \gg 2g_d R_5 > g_d R_3$, the equation simplifies to give

$$\text{CMRR} \approx 2g_m R_5.$$

For typical values $R_5 = 10\ \text{k}\Omega$ and $g_m = 5\ \text{mS}$:

$$\text{CMRR} \approx 100 \text{ (or 40 dB since it is a voltage ratio)}.$$

It is obvious from this analysis that the larger the value of R_5 the better is the common-mode rejection ratio of the amplifier.

The best solution is to use a constant current source in place of R_5. The differential amplifier with a constant current generator in place of the

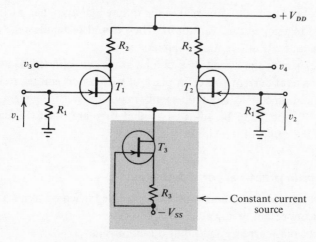

Fig. 4.26. F.e.t. amplifier with a constant current source.

resistance R_5 is shown in fig. 4.26. This arrangement can be designed to provide a CMRR of 100 dB. It is invariably used in integrated circuit differential amplifier stages where high value resistors are not economical.

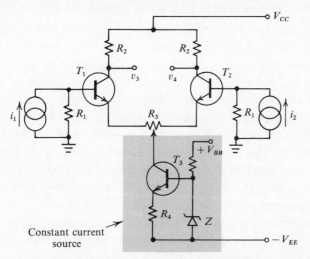

Fig. 4.27. Bipolar transistor differential amplifier with a constant current source in the emitter lead.

4.9.3 Bipolar transistor differential amplifier

Bipolar transistor differential amplifiers have the advantage that it is possible to construct more nearly identical bipolar transistors than f.e.t.s on

145

the same chip of silicon. This makes the differential amplifier particularly insensitive to temperature variations. They can also handle much larger currents than f.e.t. differential amplifiers.

Fig. 4.27 shows the circuit of a bipolar transistor differential amplifier using a constant current source in place of a common emitter resistor.

The potentiometer R_3 is a refinement which enables the currents in the transistor T_1 and T_2 to be adjusted so that they are exactly equal. This improves the CMRR of the circuit still further.

4.10 Some practical considerations

A number of practical points must be considered when op-amps are used. The most important features are as follows:

(a) input bias currents and input offset current,
(b) offset voltage,
(c) common-mode rejection ratio,
(d) slewing rate.

These are considered in the next four sections.

4.10.1 Input bias currents and input offset current

Operational amplifiers draw small currents, typically less than 1 μA, at the input terminals in order to bias the base or gate of the input transistors. Such transistors are shown in the differential amplifier in fig. 4.27 which is the type of circuit which often forms the input of integrated circuit operational amplifiers. The effect of these bias currents is to produce a spurious output voltage when there is no input signal and the output should be zero or, in the presence of an input, an additional output is superimposed on the amplified signal. Generally this extra output is undesirable and it is particularly troublesome during operation in DC conditions. It is more significant when large resistors are used in the op-amp circuitry or the overall op-amp gain is high.

The effect can be minimised by adding an extra resistance in the circuit as shown by the analysis of the amplifier circuit of fig. 4.28. Note the following parameters which apply in this analysis:

bias current I_{B1} in ($-$) terminal,
bias current I_{B2} in ($+$) terminal,
offset current $I_o = (I_{B1} - I_{B2})$,
offset output voltage V_2 and
resistance R_3 for offset balance.

146

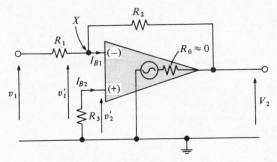

Fig. 4.28. Input bias currents.

In this analysis we assume R_i is large and R_o is small for the integrated circuit amplifier.

At node X, if the signal v_1 is ignored at present (superposition) then

$$\frac{0-v_1'}{R_1}+\frac{V_2-v_1'}{R_2} = I_{B1}$$

which gives

$$v_1' = \left(\frac{V_2}{R_2}-I_{B1}\right)\frac{R_1 R_2}{R_1+R_2}.$$

Also

$$v_2' = I_{B2}R_3$$

and the output

$$V_2 = A(v_2'-v_1').$$

For large A, from the equations above, we get

$$V_2 = I_{B1} R_2 - I_{B2}\frac{R_3}{R_1}(R_1+R_2). \tag{4.62}$$

In order to minimise the unwanted signal, V_2, the resistance R_3 is made equal to the parallel combination of R_1 and R_2 or

$$R_3 = \frac{R_1 R_2}{R_1+R_2}.$$

For this condition, (4.62) simplifies to

$$V_2 = (I_{B1}-I_{B2})\, R_2.$$

As I_{B1} and I_{B2} are nearly equal V_2 is greatly reduced. The offset current I_o is specified as $I_{B1}-I_{B2}$ and it is about 200 nA for a 741 op-amp and about 50 pA for a 740. For $R_2 = 10$ kΩ, $V_2 = 2$ mV for a 741 and 0.5 μV for a 740. It should also be noted that the offset current is subject to change with amplifier lifetime and as a result of temperature fluctuations.

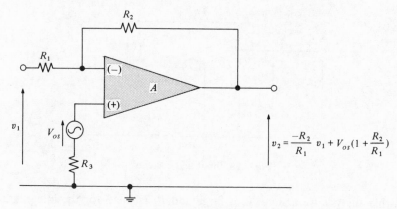

Fig. 4.29. Input offset voltage.

$$v_2 = \frac{-R_2}{R_1} v_1 + V_{os}\left(1 + \frac{R_2}{R_1}\right)$$

4.10.2 Offset voltage

The output voltage from an op-amp is not exactly zero for another reason when there is no input signal and both the $(-)$ and $(+)$ inputs are earthed: steps must be taken to set the output to zero when there is no input signal. This offset voltage arises because the transistors in the differential amplifiers are not identical and other electronic circuitry in the integrated circuits is not perfectly balanced.

The input offset voltage V_{os} is defined as the small voltage needed at the *input* to make the *output* zero. Usually the input offset voltage is a few millivolts (5–6 mV for a 741 op-amp). It can be modelled in a circuit as shown in fig. 4.29 and superimposed on the signal voltages.

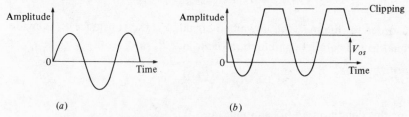

Fig. 4.30. (*a*) Input signal to inverting op-amp. (*b*) Output with offset showing clipping of the peak signal.

Thus for an input signal as shown in fig. 4.30(*a*) the output is as shown in fig. 4.30(*b*). The effect of the shift in the zero level can be a serious drawback since it limits the maximum signal that may be handled without clipping one end of the waveform.

In any practical circuit steps must be taken with special circuits to

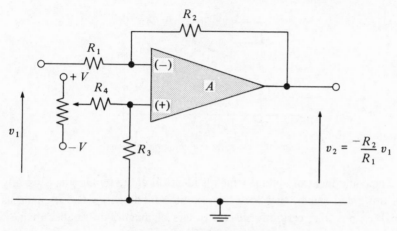

Fig. 4.31. Offset adjustment.

balance out the offset. In some op-amps connections are provided by the manufacturer to null out the offset voltage. In fig. 4.31 a typical practical op-amp circuit is shown. Both voltage offset and current offset may be corrected with this circuit.

4.10.3 Common-mode rejection ratio

We have pointed out that the two inputs on the operational amplifier make it a *differential* or difference amplifier. This type of amplifier, in the ideal case, has a large gain for the difference between the two signals at its input terminals and no gain at all (zero output) for identical input signals. In real amplifiers this latter condition cannot be obtained exactly and the amplifier always has a small but finite gain for identical input signals. This has been discussed for the case of the f.e.t. differential amplifier.

As discussed earlier, it is customary to consider a differential amplifier to have two types of gain. One is the differential gain, A, which we have used so far and the other is the common-mode gain which is defined with reference to fig. 4.32 as the ratio of the output and input voltages when the input voltage to the two terminals is exactly the same. The output from any differential amplifier with input signals v_1 and v_2 at the inverting and non-inverting inputs respectively is therefore

$$v_3 = A(v_2 - v_1) + (A_{cm}/2)(v_2 + v_1).$$

It should be noted that $A_{cm}/2$ is used here because, by definition, with $v_1 = v_2$, the output v_3 must equal $A_{cm}(v_1)$ and not $2A_{cm}(v_1)$.

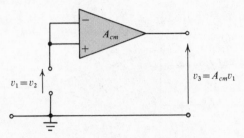

Fig. 4.32. Definition of common-mode gain.

The component of voltage which is identical at the two inputs is usually an unwanted signal, therefore the common-mode gain, A_{cm}, is made as small as possible, certainly several orders of magnitude smaller than A. (Ideally one would want $A_{cm} = 0$.) The parameter used to express this quality of an amplifier is the *common-mode rejection ratio*, CMRR, defined as A/A_{cm}. The ratio is usually expressed in decibels as $20\log_{10}(\text{CMRR})$. Typically A/A_{cm}, in a well designed amplifier, is between 10^4 to 10^5.

Example
An amplifier with $A = 10^4$ and CMRR = 80 dB has an input signal $v_2 = 0.5$mV at the non-inverting input and $v_1 = -0.5$ mV at the inverting input. A spurious signal spike of 10 mV is also picked up at both inputs. Calculate the output voltage and compare it with the output expected in the absence of the spurious signal.

Here $\qquad A = 10^4 \quad \text{and} \quad A_{cm} = 1.$

$$v_3 = A(v_2 - v_1) + A_{cm}/2(v_2 + v_1)$$
$$= 10^4(10.5 - 9.5) + 1/2(10.5 + 9.5) \text{ mV}$$
$$= 10.01 \text{ V}.$$

Ideally the output would have been 10 V therefore the error is 1 part in 10^3 or 0.1 per cent.

4.10.4 Slewing rate and full power response
The behaviour of the operational amplifier when handling large signal inputs is not the same as when handling small signal inputs at the same frequency. This is expressed by the *slewing rate* of the amplifier which is the maximum rate of change of the output voltage when the amplifier is supplying its rated output. The concept is illustrated in fig. 4.33 where the output signal obtained for a step change of input is shown. The ideal output is, of course, a step. The true output is a ramp with a time Δt to change the voltage by ΔV. The ratio $\Delta V/\Delta t$ is the slewing rate and is given in V/μs.

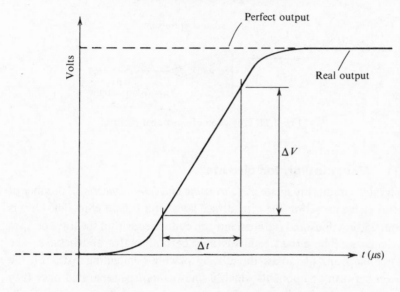

Fig. 4.33. Slew rate of the operational amplifier.

The effect is a function of the finite charging time of capacitors used in the operational amplifier's internal circuit and the load conditions. The main point to note is that when large signals are to be handled by the integrated circuit amplifier the small signal frequency response will not give a sufficient guide to the performance.

A related parameter is the *full power response frequency*. This is the maximum frequency at which the full output power may be obtained without undue distortion of the signal. Again the large signal handling ability of the integrated circuit amplifier is not as good as its small signal performance.

Precautions must be taken with certain types of integrated circuits to protect them from accidental damage. The output terminals of the μA709 must not be directly connected to ground. The usual protection is a resistance of approximately a hundred ohms in series with the output terminal. The μA741 and μA740 have this protection built in. Circuits are also protected against accidental over-voltage on the supply rails and against 'latch-up' – a condition where the amplifier output settles at an extreme of supply voltage. Details of such protection schemes may be found in manufacturer's data. The input may also be protected from too large a voltage between inputs using a circuit such as shown in fig. 4.34.

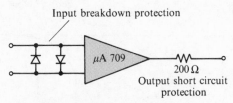

Fig. 4.34. Protection of integrated circuits.

4.11 Linear integrated circuits

Integrated circuits are made by *fabricating* and *interconnecting* a number of circuit elements within the top surface layer of a silicon chip. The chip is about 250 μm thick and the elements are contained within the top 2 or 3 μm of thickness. The circuit elements may be both active and passive. The interconnections are made directly by placing conducting paths on the silicon between components which usually cannot be separated once they have been made. The device performs *in toto* as a complex circuit and individual parts cannot be repaired. If it is damaged it is discarded as a whole.

Integrated circuits may be divided into two main types: *monolithic* and *hybrid*. Monolithic circuits are those made entirely on the surface of a single silicon chip which is called the *substrate*. This substrate usually forms an integral part of at least some of the devices used in the circuit. The interconnection between devices is made by depositing a pattern of conducting lines linking the circuit elements. The active devices are made by the usual methods of semiconductor device technology such as impurity diffusion, masking, photo-resist exposure, etc. Finally wires are bonded to contact pads at the relevant points on the circuit which require external connection and the whole device is encapsulated to form a single package.

Hybrid circuits are those where a number of separate silicon chips is used with each chip containing one or more circuit elements. These chips are mounted close together on an insulating substrate and interconnected. Finally the whole assembly is sealed off in one package.

The striking feature of integrated circuits is their small size and weight. For example a general purpose operational amplifier occupies only 5 per cent of the volume and has only 2 per cent of the weight of a comparable discrete component circuit. If the comparison is made in terms of chip size rather than package size, the reduction is even more striking.

However a much more important advantage is the much greater reliability obtained from the automatically made and much tested integrated circuit. Soldering, a notorious source of faults, is only required on a few external leads. The failure rate of the circuit as a whole is not much worse than the effective failure rate of each one of the many separate active devices used in a discrete component circuitry. Thus the failure rate should be decreased in complex electronic systems such as large computers – or an experimental device left on the moon should work for many years if made with integrated circuits.

Other advantages are that a decrease in cost is possible where large numbers of circuits of the *same* design are to be fabricated. Reduced power consumption is often obtained compared with discrete component circuits of comparable performance.

Very low drift amplifier circuits can be constructed because important components are physically very close together and change simultaneously in temperature. Circuits have been constructed with a built-in heater element to stabilise the circuit temperature. A low drift operational amplifier with a drift of less than 5 μV/week is available.

An advantage of the small size and short interconnections in integrated circuits is that spurious signals are not picked up so easily as with discrete circuits. Stray capacities to ground are reduced and faster switching is possible. In very large scale integrated circuits there may even be the possibility of a shorter propagation time for signals and a reduction of delay.

There are limitations in the power handling capacity of integrated circuits and discrete device output stages are needed to provide power output greater than about 5 watts. In almost all other respects integrated circuits can be designed to perform better than equivalent discrete device circuits.

The design of integrated circuit amplifiers does not follow the same approach as that used for discrete circuit amplifiers. The main points of difference may be summarised as follows:

(*a*) In discrete circuits the active devices are kept to a minimum for reasons of economy and reliability whereas in integrated circuits the active devices in general occupy less surface area than passive circuit elements and are therefore preferred.

(*b*) Large value resistors and capacitors, or components to small tolerances in absolute value, may be used in discrete circuits but are not available in integrated circuits.

(*c*) Matched active or passive devices are difficult to obtain and expensive for discrete circuits but are relatively easily obtained for integrated circuits.

(*d*) Inductors are not available for integrated circuits.

The amplifier circuit which takes best advantage of the points in favour of integrated circuits is the *differential* amplifier shown in fig. 4.27.

153

4.12 Worked examples

1. The signal from a test probe used for a medical investigation varies about a mean level of 10 mV amplitude and must be amplified accurately, with an operational amplifier, by a factor of 100 times before it is displayed on a chart recorder of 1 kΩ input impedance. The internal resistance of the signal source varies from 1 MΩ to 10 MΩ and the amplifier data are as follows: $A = 10^4$, $R_i = 10^5 \ \Omega$, $R_o = 100 \ \Omega$. Explain why the inverting amplifier configuration cannot be used and determine the accuracy with which the amplification can be carried out by using (*a*) a single non-inverting amplifier and (*b*) a voltage follower and a non-inverting amplifier.

Solution. The input resistance of an inverting amplifier (fig. 4.4) is $R_{in} = R_1$. $|\text{Gain}| = R_2/R_1 = 100$ from (4.2).

R_2 cannot usually be greater than 1 MΩ otherwise stray capacity affects the performance of the amplifier. Therefore, $R_{in} \approx R_1 = 10$ kΩ.

The signal source resistance is 1 MΩ–10 MΩ which is much greater than R_{in}. However in the discussion in §1.2 we showed that for efficient coupling of a voltage source to an amplifier the source resistance should be much less than R_{in}.

Therefore the inverting amplifier cannot be used. In fact the output from it in the two cases of source resistance would be ∼ 10 mV and ∼ 1 mV respectively.

(*a*) Non-inverting amplifier (fig. 4.8).

$$\text{Gain} \approx (1 + R_2/R_1) = 100 \quad \text{so} \quad R_2 = 99R_1.$$

From (4.24), input resistance $\approx \dfrac{R_i A}{1 + (R_2/R_1)} = 10^7 \ \Omega.$

For 1 MΩ source resistance R_S and $10^7 \ \Omega$ input resistance, R_{in}, the signal output

$$= v_{in} \left(\frac{R_{in}}{R_S + R_{in}} \right) \times \text{gain}$$

$$= \frac{10 \ \text{mV} \times 10}{11} \times 10^2 \approx 0.91 \ \text{V},$$

i.e. an error of ∼ 10 per cent.

For 10 MΩ source resistance the signal output

$$= \frac{10 \ \text{mV} \times 10}{20} \times 10^2 = 0.5 \ \text{V},$$

i.e. an error of ∼ 50 per cent.

The output resistance of the amplifier is

$$\frac{R_o \times 100}{10^4} = 1 \ \Omega.$$

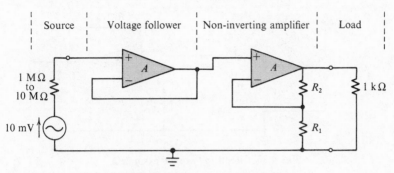

Fig. 4.35. Circuit for worked example 1.

Therefore the error in coupling to the chart recorder is ~ 0.1 per cent. Hence the total error varies from 10 per cent to 50 per cent.

(*b*) Using the arrangement shown in fig. 4.35.

Input resistance of voltage follower $= (1+A)R_i \approx 10^9 \ \Omega$. Gain $= 1$ to better than 1 part in 10^5.

[In practice this may be shunted by a common-mode resistance from the (+) input to ground.]

Now with 1 MΩ source resistance, output of voltage follower $= 10$ mV to 1 part in 10^3.

With 10 MΩ source resistance, output of voltage follower $= 10$ mV to 1 part in 10^2.

Output resistance of voltage follower $= 100/10^4 = 0.01 \ \Omega$.

Thus the error in coupling to the non-inverting amplifier is negligible.

The output resistance of the non-inverting amplifier is $R_o \times 100/10^4 = 1 \ \Omega$ and the error in coupling to the chart recorder is therefore ~ 0.1 per cent.

Hence the total error varies from 0.2 per cent to 1.1 per cent using this arrangement.

2. The circuit of a strain gauge bridge and amplifier is shown in fig. 4.36. The gauge factor (fractional change in resistance given by the strain) is 2 and the bridge resistors R_1 are each 100 Ω, the resistors R_2 are 10 kΩ and the bridge supply $v = 10$ V.

Derive an expression for the output voltage from the bridge for an incremental resistance change $\Delta R_1/R_1 = \delta$ and determine the output for a strain of 0.0005. It may be assumed that A is very large.

Solution. For the applied voltages shown in fig. 4.36:

At the node X we have

$$\frac{v - v_1}{R_1} + \frac{v_3 - v_1}{R_2} - \frac{v_1}{R_1} = 0.$$

155

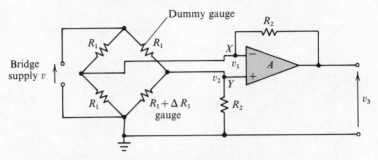

Fig. 4.36. Circuit for worked example 2.

At the node Y we have

$$\frac{v - v_2}{R_1} - \frac{v_2}{R_2} - \frac{v_2}{R_1 + \Delta R_1} = 0.$$

When A is large and $R_2 \gg R_1$

$$v_3 = v \frac{R_2}{R_1} \frac{\delta}{2 + \delta}.$$

Usually $\delta \ll 1$,

$$v_3 \approx v \frac{\delta}{2} \frac{R_2}{R_1}.$$

For the figures given $\Delta R_1 / R_1 = \text{strain} \times \text{gauge factor} = 0.001$, hence

$$v_3 = 10 \times \frac{0.001}{2} \times 100 = 0.5 \text{ V}.$$

3. The differential amplifier circuit is shown in fig. 4.37(a) fed from an input signal source v_1 with an output resistance R_1 while the other input terminal is connected to ground through a resistor R_2.

Determine the voltage output v_3 at the collector of T_1, and also the common-mode rejection ratio for the amplifier. The two transistors may be assumed to have identical values of h_{fe} and h_{ie} with h_{oe} and $h_{re} = 0$.

Solution. The equivalent circuit becomes that shown in fig. 4.37(b). Writing two loop equations we have

$$v_1 = i_1\{h_{ie} + R_1 + (1 + h_{fe})R_5\} + (1 + h_{fe})i_2 R_5,$$

$$0 = i_2\{R_2 + h_{ie} + (1 + h_{fe})R_5\} + i_1(1 + h_{fe})R_5.$$

Eliminating i_2 and using $\quad v_3 = -h_{fe}i_1 R_3$

we have

$$v_3 = \frac{-v_1 h_{fe} R_3}{R_1 + h_{ie} + (1 + h_{fe})R_5 \left\{ 1 - \frac{(1 + h_{fe})R_5}{R_2 + h_{ie} + (1 + h_{fe})R_5} \right\}}$$

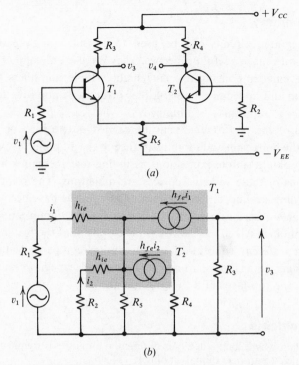

Fig. 4.37. (*a*) Differential amplifier with bipolar transistors.
(*b*) Equivalent circuit.

Usually $(1+h_{fe})R_5 \gg R_2+h_{ie}$, because we have $h_{fe} \sim 100$ and R_5 is as large as possible, consistent with reasonable power supply requirements. We may write that

$$v_3 \approx -\frac{v_1 h_{fe} R_3}{2(R_1+h_{ie})} \quad \text{if} \quad R_1 = R_2.$$

If common-mode signals are present then we have from our previous considerations in §4.9.2 that the circuit may be divided up because of its symmetry into a common-emitter amplifier with an emitter resistance $2R_5$. The method of §3.10 with an undecoupled emitter resistance gives:

$$v_3' = v_1' \frac{-h_{fe} R_3}{h_{ie}+(1+h_{fe})2R_5}.$$

Hence the

$$\text{CMRR} \approx \frac{R_5 h_{fe}}{R_1+h_{ie}}.$$

For $R_5 = 10 \text{ k}\Omega$, $h_{ie} = R_1 = 1 \text{ k}\Omega$, $h_{fe} = 200$, the CMRR ≈ 1000.

157

4.13 Summary

Linear integrated circuits can be used to construct high performance amplifiers with many useful properties. They are used in conjunction with a few external components to make highly stable amplifiers with well defined gain and impedance levels. Either inverted or non-inverted output with respect to input may be obtained.

Analysis of the performance may be carried out to high precision by including the finite gain and input resistance of the amplifier. However it is possible to obtain simple expressions assuming that the gain is very high. Applications of these versatile devices are numerous. The principal ones such as adding, scaling, differentiating and integrating have been described and chapter 6 describes oscillators using integrated circuit amplifiers.

The technology which has led to the development of these amplifiers has required a different concept of circuit design compared with discrete circuit design. Transistors, resistors, diodes and capacitors are all constructed in the top layer of a thin disc of silicon.

4.14 Problems

1. (a) Explain why a long-tailed pair differential amplifier is frequently used as the input stage to an operational amplifier.

(b) Explain why the steady-state output voltage of an operational amplifier is not exactly zero for the condition of zero input DC signal and indicate ways of overcoming this.

(c) The operational amplifiers shown in fig. 4.38(a) and 4.38(b) have very high gains and high input resistances. Estimate the gain of each stage explaining any assumptions made. Assign suitable values to R_1 in fig. 4.38(a) and R_2 in fig. 4.38(b) giving your reasons.

(d) Determine the form of feedback being used in figs. 4.38(a) and (b) and indicate the input and output impedance levels in each case.

(Birmingham University: First year)

2. The operational amplifier in the circuits of fig. 4.39(a) and (b) has an input impedance R_i between terminals 1 and 2, and an output voltage $e_0 = A(e_2 - e_1)$, where e_0, e_1, e_2 are the voltages at terminals 0, 1, 2 respectively.

Show that the amplifier output voltage in circuits (a) and (b) respectively may be written as

$$\text{(a) } e_0 = -R_1 \left(\frac{A\beta_1}{1 + A\beta_1} \right) i_s; \quad \text{(b) } e_0 = \frac{R_2 + R_3}{R_3} \left(\frac{A\beta_2}{1 + A\beta_2} \right) e_s$$

and express β_1 and β_2 in terms of the circuit parameters.

If $A = 2 \times 10^3$, $R_i = 20$ kΩ, and $R_1 = R_2 = R_3 = 10$ kΩ, find for circuit (a) the range of source impedance R_p for which $e_0/i_s = -10^4$ to within 1 per cent, and for circuit (b) the range of source impedance R_s for which $e_0/e_s = 2$ to within 1 per cent.

158

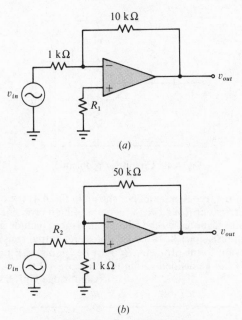

(a)

(b)

Fig. 4.38. Circuit for problem 1.

If the sources are nearly ideal (R_p very large, R_s very small) what is the frequency range for the same performance tolerance in each configuration if the open-loop frequency response of the amplifier is given by $A/(1 + 10^{-5}j\omega)$?

Comment on possible applications of these circuits.

(Cambridge University: Third year)

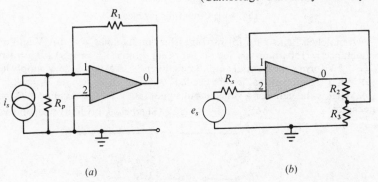

(a)   (b)

Fig. 4.39. Circuit for problem 2.

3. Write an account of the construction of linear integrated circuits. Show how the various components can be realised in a form suitable for this method of manufacture, and sketch a possible realisation of the circuit of fig. 4.40 in integrated form. Comment on how the requirement to manufacture by this method affects the problems of circuit design.

(Birmingham University: Second year Mech. Eng.)

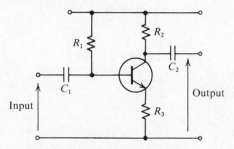

Fig. 4.40. Circuit for problem 3.

4. In the circuit of the feedback amplifier shown in fig. 4.41 the overall gain is to be defined by the resistors R_1 and R_2, which are of high precision. The specifications of the amplifier denoted by the triangular symbol include gain $-A$, where $A \geqslant 3 \times 10^4$, input resistance $R_i \geqslant 3 \times 10^5$ ohms and output resistance negligible. For reasons irrelevant to the problem, R_2 must be 10^5 ohms. Find the maximum overall gain which can be defined within tolerance of $+0$ to -0.1 per cent.

(Cambridge University: First year)

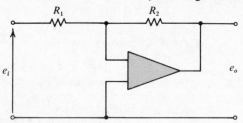

Fig. 4.41. Circuit for problem 4.

5. Two matched f.e.t.s with forward transfer conductance = 5 mA/V and drain conductance = 150 μmhos when at working point $V_{ds} = +5$ V, $I_d = 1$ mA, $V_{gs} = -2$ V form a differential amplifier between $+25$ V and -25 V supply lines.
   (a) What component values are needed? Assume $R_3 = R_4$ and $R_1 = R_2$.
   (b) What is the gain for a single input signal (i.e. $v_2 = 0$)? Assume $1/R_5 = G_5$ is negligible.

(Sheffield University: Second year)

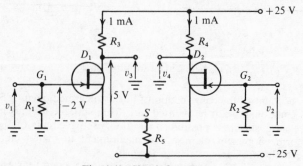

Fig. 4.42. Circuit for problem 5.

6. Outline the desirable properties of an operational amplifier for use in analogue computation. Fig. 4.43 shows a simple operational amplifier: comment on the general properties of the circuit in the context of the desirable properties of an operational amplifier. In fig. 4.43 the input transistors are to operate with $I_c \simeq 0.5$ mA, and the output is required to be at least ± 10 V across an external load of 5 kΩ. Also the dissipation in T_3 must not exceed 120 mW at zero output. Assuming that V_{be} for T_3 is about 0.6 V, choose suitable resistance values for the circuit and justify your choice, indicating any assumptions made. Using the relation $r_{in} \simeq 25\beta_0/I_c$, and assuming that $\beta_0 = 50$ for all transistors, estimate the amplifier gain for small output signals into the 5 kΩ load.

(Imperial College: Second year)

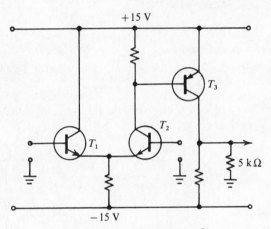

Fig. 4.43. Circuit for problem 6.

Quiz 2 (chapters 3 and 4)

Underline the correct statements.

1. The symbol shown in fig. 1 is that of a (a) npn transistor, (b) pnp transistor, (c) n-channel f.e.t., (d) MOST.

2. The ratio i_c/i_b in a bipolar transistor is the (a) alpha, (b) beta, (c) g_m, (d) h_{ie}, (e) h_{fe}, (f) h_{oe}, (g) h_{re}.

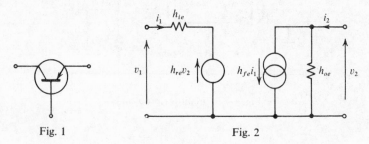

Fig. 1   Fig. 2

3. A good transistor should have (a) a high leakage current, (b) a high maximum value for V_{CE}, (c) a low knee voltage, (d) a low h_{fe}.

4. The equivalent circuit for the bipolar transistor shown in fig. 2 is called the (a) T-equivalent network, (b) hybrid-parameter network, (c) hybrid-π network.

5. The equivalent circuit of fig. 2 may be used to represent the transistor when (a) the signals are small, (b) the signal frequency is very high, (c) the signal frequency is low, (d) the signals are large.

6. In fig. 2 the dimensions of h_{oe} are (a) dimensionless, (b) ohms, (c) siemens, (d) amps, (e) mhos, (f) mA/volt.

7. The common-emitter leakage current of a bipolar transistor is high if (a) temperature is high, (b) V_{CE} is high, (c) h_{fe} is high, (d) frequency is high.

8. The circuit of fig. 3 is (a) an inverting amplifier, (b) a differential amplifier, (c) a non-inverting amplifier, (d) a voltage follower.

162

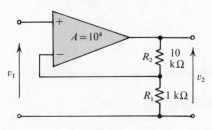

Fig. 3

9. The voltage gain of the circuit of fig. 3 is (*a*) 10, (*b*) 11, (*c*) −10, (*d*) 1, (*e*) 10^4, (*f*) −11.

10. The amplifier shown in fig. 4 when compared with the one in fig. 3 has (*a*) higher gain, (*b*) inverted output, (*c*) higher input resistance, (*d*) lower gain.

11. The circuit of fig. 3 may be made into a voltage follower by making (*a*) $R_2 = 0\ \Omega$ and $R_1 = \infty$, (*b*) $R_1 = 0\ \Omega$ and $R_2 = \infty$, (*c*) $R_1 = R_2 = 1\ \text{k}\Omega$.

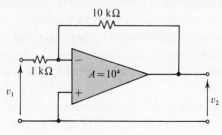

Fig. 4

163

5

Negative feedback

5.1 Introduction

Feedback occurs where some part of the output of a system also appears back at its input and so modifies the input signal. It occurs in *every* system. If the feedback is undesired, then the input and output circuits of a system must be well separated and screened. Remember that if the circuit is meant to handle signal frequencies of a few megahertz, which are in the broadcast band, these will be readily radiated from the output circuit wiring and could easily be picked up by the input wiring. Also, if the input and output stages share a common power supply, then care must be taken to 'decouple' this as far as signal frequencies are concerned. Otherwise feedback could occur.

This chapter describes intentional feedback, which, when it is properly applied, can improve almost every performance feature of an amplifier. It can widen its frequency response, reduce the effects of component ageing, microphony and hum pickup, and stabilise the overall gain closely to some figure required by a designer. These are a few of its benefits which are more fully explained in the following sections. The subject is important because virtually every good quality or precision amplifier made today is likely to use feedback.

There is one troublesome effect of feedback on amplifier performance; it can create a tendency towards instability if the system is poorly designed. However, feedback which is specifically designed to make oscillators or switches is useful in its own right and it is described in chapter 6.

5.2 Definitions of feedback

A feedback circuit is one where part of the output signal is added to the input. It can be added so as

(*a*) to reduce the input – such feedback is then called inverse, degenerative or negative feedback, or

164

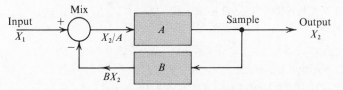

Fig. 5.1. A system with feedback.

(*b*) to increase the input – such feedback is then called regenerative or positive feedback (see chapter 6).

Fig. 5.1 shows a system, which need not necessarily be an electrical one, with some input signal, X_1, applied to it and we wish to define how the output signal, X_2, is related to X_1. If the block marked A is an amplifier of gain A, then for an output signal X_2 the required input is X_2/A. The amplifier is assumed to be unilateral and unaffected by the feedback circuit, B, which *samples* its output. In other words, the feedback network B takes a negligible signal from the output of the amplifier A. The signal X_2 applied to the feedback circuit, B gives a signal BX_2 to be *mixed* with the input. The arrows leading to the blocks are important in deciding which is the input and which the output of each block. Then A and B can be simply treated as 'gains'.

At the input to the system, the signals are mixed (the ways in which this can be done practically are described in §5.3). The mixer shown in fig. 5.1 gives an output equal to the difference of the signals fed to it, so that

$$X_2/A = X_1 - BX_2. \tag{5.1}$$

By re-arranging, we get

$$\frac{X_2}{X_1} = \frac{A}{1+AB} = \frac{\text{forward gain}}{1+\text{loop gain}} \tag{5.2}$$

$$= K,$$

the overall gain of the system. The words 'loop gain' refer to the gain of a signal path forward through the amplifier A and then through the feedback path B in a closed loop back to its origin. The gain round this loop is AB. The size and sign of AB, the loop gain, very much controls the usefulness of the feedback.

Let us consider some possible conditions for A and B and see how they affect the overall gain, K.

Case (*a*). With no feedback, so $B = 0$,

$$K = \frac{A}{1+0} = A \quad \text{as expected.}$$

Case (*b*). If either *A* or *B* is negative, so $(1 + AB)$ is less than unity,

$$K = \frac{A}{\text{small quantity}} \gg A.$$

This is positive feedback.

Case (*c*). If the loop gain, *AB*, is exactly equal to -1,

$$K = \frac{A}{1 - 1} = \text{infinity}.$$

This is a circuit whose gain tends to infinity and so it is of no use as an amplifier. It will normally generate an output in the absence of an input, and so be an oscillator (see chapter 6).

Case (*d*). If *A* and *B* are both positive or both negative, and the magnitude of *AB* is much greater than unity,

$$K \approx \frac{A}{AB} = \frac{1}{B}. \tag{5.3}$$

This is a circuit whose gain is no longer defined by the actual gain of the internal amplifier in the forward path, but only by the feedback path. If the feedback path is constructed of passive components, and if these are chosen to be very stable types, then the amplifier overall gain, *K* can also be very stable. This is an extremely useful property of feedback.

As an example, consider a circuit where $B = \frac{1}{100}$ and is made from passive components; *A* is of the order of 10^4 but may be variable, then the gain with feedback, *K*, is given by

$$K = \frac{A}{1 + AB} = \frac{10^4}{1 + 10^2} = 99,$$

which is close to $1/B$. If the amplifier happened to be constructed of active devices whose gains were all higher than expected, so that $A = 10^5$, then

$$K = \frac{10^5}{1 + 10^3} = 99.9.$$

For a rise in *A* of 1000 per cent, *K* has changed by about 1 per cent. Thus the stability of the forward gain, *A*, does not matter very much providing that *AB* is high. This is explored further in §5.4.

5.3 Realisation of A and B

The way in which the feedback network samples the output of an amplifier and mixes the signals at the input depends on whether the overall circuit is

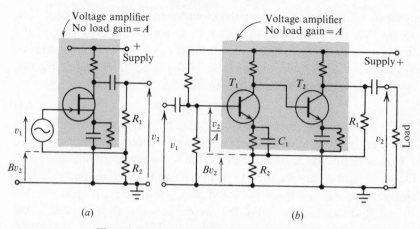

Fig. 5.2. Voltage amplifiers with voltage feedback.

required as a voltage or current amplifier or some other type. Here, only a few of the many possibilities are explained but these should show the methods involved.

5.3.1 Voltage feedback applied to a voltage amplifier

Considering first a voltage amplifier, whose gain A may be the voltage gain of one or several stages of valves or transistors: then the feedback circuit should pass a fraction B of the output voltage to the input to make a better voltage amplifier. Ways of doing this are shown in fig. 5.2(a) and (b).

The feedback circuit consists of two resistors, R_1 and R_2, in each amplifier. R_1 is joined to the amplifier output terminal but is assumed to be large enough not to load it appreciably. This could be achieved by choosing it so that the current passing through R_1 and R_2 to the signal-common line is of the order of one tenth of the current in the transistor. The input signal mixing is most clearly shown in fig. 5.2(a). The input to the f.e.t. gate is clearly the sum of the generator signal and Bv_2. As the gate current is so small, R_1 and R_2 will act as an unloaded potentiometer and B will be $R_2/(R_1 + R_2)$. The circuit has the disadvantage that the generator must be 'floating', that is, neither lead is connected to ground. Thus a laboratory oscillator connected as the source v_1 in fig. 5.2(a) could short out the resistor R_2 if one of its output terminals was an earth. However a microphone, say, can be a 'floating' source if its earthed screening is kept separate from the two signal wires.

A complicated but more commonly met arrangement is shown in fig. 5.2(b). Sampling of the output is achieved through R_1 and the other feedback resistor, R_2, is connected between the input device, T_1, and the

167

signal-common line. Comparing this circuit with that shown in fig. 3.7(*a*), we see that the decoupling capacitor C_1 is put across only part of the emitter resistor of T_1 and leaves an amount equal to the desired value of R_2 un-decoupled. By allowing a signal of Bv_2 to be developed across this resistor, the voltages shown at the input to fig. 5.2 give:

$$v_1 = v_2/A + Bv_2. \qquad (5.4)$$

This is the same as (5.1). Therefore this circuit directly gives the right mixing at the input so that the signal fed back is *subtracted* from the input.

Now how are the resistor values chosen to achieve a given value of B? Ideally R_1 and R_2 are required to form a potentiometer across the output voltage v_2 so that a voltage Bv_2 is fed back. However, it is apparent that R_1 and R_2 do *not* form an unloaded potentiometer as R_2 is shunted by the impedance to earth looking into the emitter of the transistor. If this is r, then

$$B = \frac{R_2 \| r}{R_1 + R_2 \| r} \quad \text{(where } \| \text{ means 'in parallel').} \qquad (5.5)$$

r can either be found either by network analysis or can be approximated to either $1/g_m$ or h_{ie}/h_{fe} for f.e.t. or bipolar devices respectively used as T_1. Since it is desirable to have B defined by R_1 and R_2 and not by the device parameters which are variable, R_2 is normally selected to be much smaller than r where possible.

Case (*d*) in the previous section required that A and B be either *both* positive or *both* negative to get a useful amplifier combination. Since the resistors R_1 and R_2 in the feedback path do not invert the signal, B is a positive number (albeit less than unity); and so A will be required to be a positive gain, i.e. to have *no* phase change. This amplifier could be made of two bipolar transistors in the common-emitter connection (each has 180° phase change from input to output), or a f.e.t. followed by a bipolar transistor as shown in fig. 3.20. A four-stage amplifier might appear attractive because the gain would be higher but is seldom used as unwanted phase shifts at high frequency will give the possibility of instability (see §5.9).

5.3.2 Voltage feedback applied to differential amplifier

In chapter 4, it has been pointed out that integrated circuit amplifiers are cheap, very reliable, and easy to build into systems because they only have a few external leads to connect. Most of these amplifiers have differential inputs (see chapter 4 and fig. 4.1) so their no-load voltage gain, A, and input voltages and output voltage v_2 are related by

$$v_2 = A \text{ (difference in the input voltages).} \qquad (5.6)$$

The input connections are shown joined to points marked $(+)$ and $(-)$ on

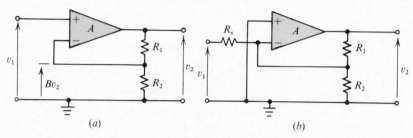

Fig. 5.3. (a) Volatge feedback on a differential amplifier. (b) Feedback *and* signal input applied to the inverting input of a differential amplifier.

the amplifier block, and that joined to the ($-$) point is called the 'inverting input'. It is the *only input* to which feedback is normally applied through passive resistors to drop the gain of the amplifier. (Chapter 6 explains what happens when feed back is applied to the other input.)

However the signal can be applied to either input. Fig. 5.3(a) shows the arrangement where the signal v_1 is applied to the non-inverting input. The feedback defined by resistors R_1 and R_2 makes the inverting input signal equal to Bv_2 so (5.6) is

$$v_2 = A(v_1 - Bv_2),$$

whence

$$\frac{v_2}{v_1} = \frac{A}{1 + AB}. \tag{5.7}$$

The feedback fraction B is very closely $R_2/(R_1 + R_2)$ since the current taken at the input of the amplifier is usually small. (The effect of a finite input resistance for the amplifier was explored in chapter 4.)

The alternative arrangement uses the inverting input to the amplifier both for feedback and for the applied signal v_1 and it becomes the familiar operational amplifier if R_2 tends to infinity. Then the circuit gain closely approaches $-R_1/R_s$ (see (4.2)) where R_s is the resistance in the lead from the signal source and includes the source resistance. If the source does not have a well defined resistance, then the overall voltage gain of the amplifier will not be well defined unless the source resistance is low. This circuit is really rather more complicated and a more fruitful analysis is done by considering what currents appear at the input terminal for given output voltages. Further analysis will be done on this type of amplifier in §5.7.3 where its uses are also explained.

5.3.3 Current feedback applied to a current amplifier

Fig. 5.4 shows a current amplifier such that an input current i_3 on the left of the dotted block representing an amplifier gives an output current Ai_3

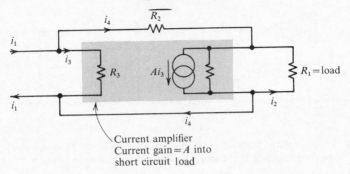

Fig. 5.4. Current feedback on a current amplifier.

from the generator on the right of the block. Let this amplifier cause an output current i_2 to flow in the load R_1.

Also shown in fig. 5.4 is a feedback path which allows a current i_4 to flow clockwise round the dotted amplifier back to the input terminals and then to return through R_2 to the output again. The polarity of the current generator Ai_3 will tend to give rise to currents i_2 and i_4 in the directions shown. By considering either of the input terminals as a node, we have the input current i_1 to the overall circuit given as

$$i_1 = i_3 + i_4. \tag{5.8}$$

Thus for a given signal input, i_1, i_3 will be reduced if i_4 is allowed to flow and the output, which is related to Ai_3, will be reduced too. Thus the feedback reduces the output for a given input and hence reduces the gain. This feedback is, by definition, negative feedback. Note that this has been realised with a *phase-reversing* amplifier which is apparent from the direction given to the output generator, Ai_3.

We may write an equation showing that the voltages round the loop going clockwise must sum to zero; so

$$-i_3 R_3 + i_4 R_2 - i_2 R_1 = 0. \tag{5.9}$$

Remembering that a good current amplifier is one which has *low* input resistance, R_3; i.e. it accepts almost all the current offered to it; and also that when the gain is high i_3 will be small, we can ignore the first term in (5.9) to give

$$\frac{R_1}{R_2} \approx \frac{i_4}{i_2} = B. \tag{5.10}$$

Thus we have defined B by ordinary resistors as the ratio of the current fed back to that appearing in the load. Note that one of the resistances is that of the load itself and that if this is not well defined, then B will not be well defined either. An example of such feedback is where a single bipolar

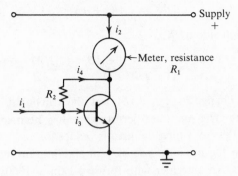

Fig. 5.5. Current feedback in an amplifier used to increase
a meter's sensitivity.

transistor is used to increase the sensitivity of a current meter as shown in fig. 5.5. If the hybrid parameter model is used for the transistor, then the circuit becomes that of fig. 5.4, where $h_{ie} = R_3$ and $h_{fe} = A$ (if h_{re} is ignored).

These amplifiers and feedback networks are just a few of the types that can be built. It is helpful to be able to examine a circuit to see if these two parts can be recognised. Many features of the performance of the whole amplifier can then be speedily determined. However exactly the same answers can be obtained by doing an ordinary network analysis on the amplifier such as that done in chapter 2 for the source follower and in chapter 3 for the emitter follower amplifier. In both circuits, *all* of the output voltage appeared back at the input, so B was unity and the circuit gain which should approximate to $1/B = 1$ was found to be true!

We have seen in the expression for the device gain with feedback, $A/(1 + AB)$, that we have *lowered* the gain by the factor $(1 + AB)$. Thus we would need more amplifier stages to amplify a given small signal by a certain fixed factor. There must be clear benefits to make feedback worthwhile; these are described in §§5.4 to 5.8.

5.4 Stabilisation of gain

Say that A, the gain of an amplifier without feedback, varies due to changes of supply voltage or circuit ambient temperature.

Then how much does K, the gain with feedback vary? If we differentiate the expression $K = A/(1 + AB)$ with respect to A, we get

$$\frac{\mathrm{d}K}{\mathrm{d}A} = \frac{(1 + AB)1 - AB}{(1 + AB)^2} = \frac{1}{(1 + AB)^2},$$ (5.11)

171

so $$dK = dA/(1+AB)^2$$

and dividing each side by K,

$$\frac{dK}{K} = \frac{dA}{(1+AB)^2}\frac{1+AB}{A} = \frac{(dA/A)}{1+AB}. \tag{5.12}$$

The fractional change in A is dA/A but the fractional change in K is $(1+AB)$ smaller. The amplifier performance has been improved by the same factor as that by which the gain is reduced.

To emphasise the practical use of gain stabilisation, consider the following example. An amplifier with forward gain = 1000 has a feedback path of gain 0.2 (i.e. $\frac{1}{5}$) in order to stabilise the overall gain to about 5. How much will this alter if the amplifier foward gain drops by 10 per cent?

From (5.12) the change in overall gain is,

$$\frac{dK}{K} = 10 \text{ per cent} \times \left(\frac{1}{1+1000\times0.2}\right) = 0.05 \text{ per cent.}$$

This is the sort of stability which would be suitable for many instrumentation uses. Note that a high, but relatively unstable gain has been traded for a lower gain and good stability. The former is easily available with present integrated circuits of the 709, 741 series having gains of 10^4 or 10^5 respectively (see chapter 4).

5.5 Increase in frequency response

We can represent an amplifier whose gain, A, starts to fall at some high frequency by writing the forward gain as $A/(1+j\omega\tau_1)$. (B is assumed independent of ω.) This was shown graphically in the Bode plot of |gain| and phase angle against frequency in chapter 1, fig. 1.22.

The expression for the overall gain K with feedback then is

$$K = \frac{A}{1+j\omega\tau_1}\bigg/\left(1+\frac{A}{1+j\omega\tau_1}B\right) = \frac{A}{1+AB+j\omega\tau_1}$$

$$= \frac{A/(1+AB)}{1+j\omega\tau_1/(1+AB)} = \frac{K}{1+j\omega\tau_1/(1+AB)}, \tag{5.13}$$

so the effect of the imaginary term is *reduced* by $(1+AB)$.

The frequency at which the gain drops to about 0.7 of its mid-frequency value is given when the real and imaginary parts of the denominator of the gain expression are equal, so this new turnover frequency is $(1+AB)/\tau_1$ rad/s. This is a higher frequency than before feedback by the factor $(1+AB)$. (Could you have guessed?!)

172

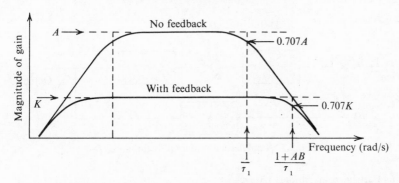

Fig. 5.6. Improvement in bandwidth given by feedback.

The improvement is shown graphically in fig. 5.6. With feedback the gain stays constant to a higher frequency. A similar analysis can be done for the lower turnover frequency and a similar improvement can be shown to occur there. Note that this is an approximate relationship. It has been assumed that the amplifier has only one main time constant τ_1 when in reality if it is a multistage circuit, it may have several in the same frequency range. This will much reduce the improvement indicated here *and* could give instability problems which will be mentioned later.

It should be noted that the frequency response is widened by the factor $(1 + AB)$ and of course the gain with feedback is $(1 + AB)$ less than without it. Thus the relation can be remembered,

$$\text{Gain} \times \text{bandwidth} = \text{constant.}$$

This is only approximately true if the amplifier has several time constants.

5.6 Reduction of the effect of internal disturbances in the amplifier

The general term 'disturbance' can be taken to mean any unwanted signal in the amplifier; it can be from poorly filtered supplies, thermal e.m.f.s in the circuit, Johnson noise in the circuit, or even non-linearity in the amplifier when the output is no longer directly related to the input. In each case, the unwanted effect is included as a disturbance, D, shown in fig. 5.7. To keep the analysis general, the disturbance is shown occurring some way between the input and output of the amplifier. The total forward gain is now $A_1 A_2$.

We can now write equations for the output signal X_2 in terms of the other signals shown on fig. 5.7:

$$X_2 = A_2(\text{output of } A_1 + D)$$
$$= A_2\{A_1(X_1 - BX_2) + D\}.$$

173

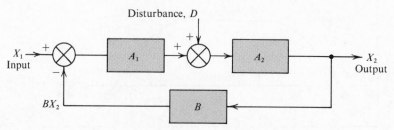

Fig. 5.7. Amplifier with internal disturbance, D.

This can be re-arranged to give

$$X_2 = X_1 \left(\frac{A_1 A_2}{1 + A_1 A_2 B}\right) + \frac{DA_2}{1 + A_1 A_2 B}. \qquad (5.14)$$

The first part of this equation gives [output = input × gain], remembering that the forward gain is now $A_1 A_2$. The second part of the equation shows the disturbance DA_2 reduced by the factor $(1 + \text{loop gain})$. Now DA_2 is the disturbance that would have reached the output without feedback and this has clearly been reduced by a useful factor in a feedback amplifier where the loop gain is large. When $A_1 A_2 B \gg 1$, (5.14) gives: $X_2 = X_1(1/B) + D(1/A_1 B)$. Now the disturbance is divided by $A_1 B$ whereas the wanted signal is only divided by B. Thus the signal has been improved relative to the disturbance. This has only been possible because A_1 is noise free amplification.

In the case of noise from random fluctuations, the noise that is developed in the input to the first stage of the amplifier is most important as it is multiplied by the whole gain of the forward amplifiers and appears at the output as $DA_1 A_2$. It is reduced by the factor $(1 + \text{loop gain})$ but as the gain is also dropped by this factor, more stages are needed to amplify given small signals. An improved *signal to noise* ratio can only be obtained in this case by choosing low noise devices for the first stage of the amplifier.

5.7 Effect on the input and output impedances of the amplifier

Feedback can be applied in several ways. It can sample output current *or* output voltage and feed back a voltage *or* a current in parallel *or* in series with the input. It is beyond the scope of this book to explore all of these possibilities. Generally the effect of feedback will be to *improve* the type of amplifier by the factor $(1 + AB)$. If output current is sensed, then the amplifier will have a better *current handling* output stage; i.e. it will have a *high* output impedance. To take another example, if the signal appearing back

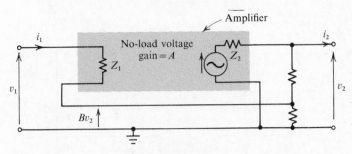

Fig. 5.8. Voltage amplifier with voltage feedback.

at the input is a *voltage* proportional to some variable of the output, then the input circuit will be a better *voltage* handling stage; i.e. one with *high* input resistance.

5.7.1 Voltage amplifier with voltage feedback

A voltage amplifier with voltage feedback is shown in fig. 5.8. The tinted block encloses an amplifier model of no-load voltage gain A, input impedance Z_1 and output impedance Z_2. Resistors are connected to 'sample' the output voltage v_2 and to feed back a fraction Bv_2 to the input circuit as shown, and only a very small current flows in them.

To calculate the input impedance of the amplifier, let us assume that the current i_2 from the output is small, so that the voltage drop across Z_2 is negligible. Then we can write the voltage at the output, v_2 as

$$v_2 = Kv_1 = v_1 A/(1+AB). \tag{5.15}$$

Now the current i_1 flowing into the amplifier can be written as the voltage across Z_1 divided by Z_1. Hence

$$i_1 = \frac{v_1 - Bv_2}{Z_1} = \frac{v_1}{Z_1}\left(1 - \frac{AB}{(1+AB)}\right). \tag{5.16}$$

Since the input impedance of any circuit is the voltage applied to it divided by the current flowing in, then from (5.16) we can write the input impedance as,
$$\frac{v_1}{i_1} = \frac{Z_1}{1 - AB/(1+AB)} = Z_1(1+AB). \tag{5.17}$$

At midband frequencies when AB is a large positive number, the impedance is increased. Thus the amplifier is an improved voltage amplifier as it is less likely to load, i.e. to drop the voltage at the terminals of any voltage source from which it is supplied.

To determine the output impedance of the amplifier with feedback, let a current i_2 be drawn from the output by some load which is not shown.

175

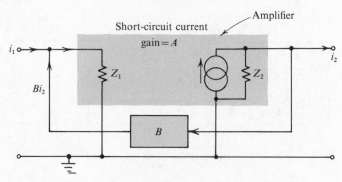

Fig. 5.9. Current amplifier with current feedback.

Assuming that the feedback components draw negligible current compared with i_2, we can write that the signal generated in the amplifier output circuit must be $v_2 + Z_2 i_2$. If the no-load gain of the amplifier is A, then

$$\text{Voltage input to tinted amplifier block} = (v_2 + Z_2 i_2)/A. \quad (5.18)$$

So the overall input v_1 is given as

$$v_1 = \{(v_2 + Z_2 i_2)/A\} + B v_2,$$

or re-arranging

$$v_2 = v_1 \left(\frac{A}{1+AB}\right) - i_2 \left(\frac{Z_2}{1+AB}\right)$$

$$= v_1(\text{no-load gain}) - i_2(\text{output impedance}). \quad (5.19)$$

So the output impedance is *reduced* by the factor $1+AB$ and the circuit is a better voltage amplifier as there will be a smaller voltage drop in the output circuit resistance when a current is drawn by a load.

5.7.2 Current amplifier with current feedback

A current amplifier with current feedback is shown schematically in fig. 5.9. The tinted block represents an amplifier with a current gain A, input impedance Z_1 and output impedance Z_2, or admittance of $1/Z_2$. A network samples the output current and feeds back a fraction $B i_2$ to the input. If the current into the feedback network, B, and through Z_2 is much less than i_2, the output of the current generator is i_2 also and thus the current into the amplifier is i_2/A. Hence we can write the input voltage, v_1 as

$$v_1 = Z_1 \times i_2/A$$

and the current at the input terminal, i_1, is given by

$$i_1 = i_2/A - B i_2. \quad (5.20)$$

176

So the impedance with feedback is given by

$$\frac{v_1}{i_1} = \frac{Z_1}{A\{(1/A) - B\}} = \frac{Z_1}{(1 - AB)}. \qquad (5.21)$$

The first thing that will be noticed about this expression is that the denominator is $1 - AB$ whereas before it has always been $1 + AB$. This is because positive feedback will occur if both A and B are positive. This can be checked by examining (5.20). By definition, for negative feedback, the term Bi_2 should be such as to reduce the amplifier input and hence its output. This can be seen clearly if (5.20) is rewritten as

Amplifier input, $i_2/A = i_1 + Bi_2$.

Thus a sign change is wanted either in A or in B to make the feedback negative. In fig. 5.4, it was assumed that A included a sign change and the normal expression for loop gain resulted. If the amplifier is a differential amplifier, then the inverting input can be chosen as that to which one applies the feedback. Alternatively, if the amplifier is made up of sign-inverting amplifying stages, then an odd number of stages is wanted.

It should be mentioned that many electronics textbooks use the expression $A/(1 - AB)$ for the amplifier gain with feedback, whereas books on control systems always assume that the feedback can be inverted as it is mixed back at the input. The convention used in electronics textbooks results in one sign in the algebra being different throughout but the meaning is identical, as either A or B must be negative.

The input impedance which was calculated in (5.21) is lower than that for no feedback by the factor $(1 + |\text{Loop gain}|)$. That is, the circuit will accept more of the current flowing into it from a given source and so be a better current amplifier.

A similar method to that described at the end of the last section will prove that the output impedance is changed too. As expected, it is changed by the factor $(1 + |\text{loop gain}|)$ and the feedback *increases* it; i.e. less of the output current is wasted in the output resistance and the circuit is a better current amplifier.

†5.7.3 Trans-conductance or trans-resistance amplifiers

A trans-conductance amplifier is one where the gain is best expressed as a conductance; that is, the output *current* has some well defined relation to the input *voltage*. The f.e.t. basically has such a gain mechanism and we may desire to apply feedback to make its *conductance* more stable and well defined. Such circuits may be used to drive the deflection system of a tele-

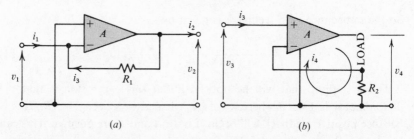

Fig. 5.10. The use of feedback across an integrated circuit amplifier to define its gain as: (*a*) Trans-resistances, v_2/i_1, (*b*) Trans-conductance, i_4/v_3.

vision set. Most cathode-ray tubes need current inputs to the deflection coils to move the spot. Alternatively, we can make a good voltage sensing circuit by connecting the output of a conductance amplifier to a simple ammeter used to display the reading.

Similarly trans-resistance amplifiers have a *voltage* output well defined in terms of the *current* input. Such circuits are extensively used, as the first amplifier stage after a television camera tube, some types of light meter elements, or any other device which gives better linearity and performance when its output current is sensed rather than its output voltage.

Basically, the way to improve such amplifiers can be explained by considering (5.3). This stated that the gain with feedback tends to $1/B$ if the loop gain, AB, is much greater than unity. So far we have considered B to be a dimensionless quantity. However if the feedback path is made up of a component or components where the output *voltage* is sensed and a *current* proportional to it is mixed with the input, then B will have the dimensions of a conductance. So the overall amplifier will tend to be the reciprocal of this, a trans-resistance.

Consider the circuit shown in fig. 5.10(*a*). The feedback component is clearly R_1. The circuit shows that a current i_3 flows out of it to the amplifier input and this current is mixed with the input current i_1 as they are joined to the same point or node. The output parameter that is sensed is the voltage v_2, irrespective of the value of i_2 the output current. Therefore the *output* circuit is improved by feedback to be better at voltage handling; i.e. the output resistance is lowered by feedback. At the input, because *current* is fed back and mixed with input currents, feedback will improve the amplifier as a current accepting circuit and give it a lowered input resistance by the factor $(1 + \text{loop gain})$. The proof of this can be found in the more advanced textbooks listed in appendix A, but the method is similar to that developed in the equations (5.16) to (5.19).

More advanced students may care to examine the performance of the summing operational amplifier (§4.7.2) as an example of the circuit of fig. 5.10(a). There the input current i_1 is the sum of the input currents coming from as many sources as there are input resistors.

The circuit of fig. 5.10(b) again has one feedback component. Here it is clearly the output current, i_4, that is sensed as it flows through R_2, and a voltage $i_4 R_2$ is fed back to the inverting input. This circuit is a dual of circuit fig. 5.10(a). The gain that is stabilised is the ratio of output current to input voltage and so the gain is a conductance. Since it is the output current that is sensed, the circuit works as a better source of output current and has a raised output impedance compared with the normal integrated circuit amplifier. Also the signal mixed back at the input is a feedback voltage and so the circuit will operate as a better voltage accepting circuit and have a higher input resistance than a normal circuit without feedback.

An interesting amplifier results on connecting the output of fig. 5.10(a) to the input of fig. 5.10(b). Because the first stage has (output voltage)/ (input current) well defined and the second stage has (output current)/ (input voltage) well defined, the overall amplifier is a good current amplifier of gain R_1/R_2. Note that the gain of the first stage is R_1 which is a trans-resistance, because the gain will tend to $1/B$ from (5.3), and B is $1/R_1$. Similarly, the gain of the second stage is $1/R_2$, a trans-conductance. The coupling between the stages is good because feedback has improved both the output of fig. 5.10(a) and the input of fig. 5.10(b) as voltage handling stages; so there is only a very small loss of gain here and the overall gain is almost exactly the product of the gain of each stage. The advantage of this current amplifier is that a very high gain can be obtained from the two stages and yet the performance is controlled by only two resistors. If these are of the high stability types (wanted for good instrumentation) whose values will not change with age or temperature, then they are expensive. So this method gives a cheaper solution than a two-stage amplifier with each stage having two resistors defining the feedback.

If each of the amplifiers is not a high gain integrated circuit amplifier but just one device, then, because the gain per stage is lower, the idealised results predicted by feedback may not be so accurately achieved. But the circuit of fig. 5.11 can be seen to be the same as fig. 5.10(a) followed by fig. 5.10(b) as far as the feedback components R_1 and R_2 are concerned. Thus the current gain, i_4/i_1, will be given approximately by R_1/R_2 again. The circuit of fig. 5.11 does not need extra components if the DC working points of the transistors are carefully chosen so that the collector voltage of the first transistor is equal to the bias required by the second transistor added to the voltage drop across R_2. Advanced students might like to

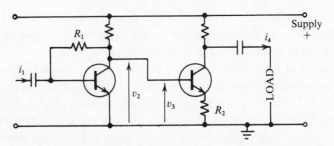

Fig. 5.11. A two-stage bipolar transistor amplifier equivalent to the circuit of fig. 5.10(*a*) followed by that of fig. 5.10(*b*); overall it is a good current amplifier.

pursue the topic of how to select components for a current gain of, say, 100. Much of the design will have to be done by trial and error but it could be made easier by adding in series with R_2 a decoupled resistor to make the selection of the DC working point more flexible.

It may appear puzzling that bipolar transistors which are basically good current handling stages can be made to have good voltage coupling between them. This is the effect of feedback if it is applied in considerable quantities – the basic amplifier is changed in performance to be almost entirely dependent on what is fed back and how this is done. Advanced readers may care to consider how the two stages of fig. 5.11, if connected in the opposite order, would give a good voltage amplifier of gain R_1/R_2.

The performance of all these circuits can be solved by network analysis such as that used in chapter 4. Then the accurate expressions for gain and the errors caused by finite values of voltage or current gain, input resistance and output resistance can be obtained. This is a much more lengthy method than obtaining the results by the methods shown in this chapter by adopting a 'systems approach'. But the systems approach has made assumptions about 'sampling', 'mixing' and high loop gain which should be checked for their validity if the results are to be accurate.

†5.8 Factors for and against feedback over several stages

When we need an amplifier of several stages to get a high forward gain A before feedback, then we have a choice of how we may apply feedback to improve the amplifier's performance. Either we can apply some feedback across each stage or we can put it in one loop across the whole amplifier. Are there benefits which will make us choose one course or the other?

Consider the amplifiers shown in fig. 5.12. Arrangement (*a*) is a multistage amplifier with each stage having a forward gain of A. B_1 is the feedback required on each stage to get an overall gain K_1. The other possi-

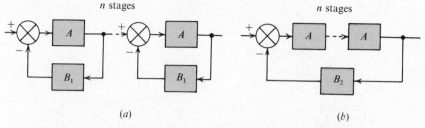

Fig. 5.12. Multistage amplifier with: (a) feedback on each stage,
(b) overall feedback.

bility, (b), is to put all the amplifiers in cascade when they will have a gain
of A^n and then to apply some different overall feedback B_2, to get the
overall gain, K_2. We can write the gains K_1 and K_2 as:

$$K_1 = \left(\frac{A}{1+AB_1}\right)^n \quad \text{and} \quad K_2 = \frac{A^n}{1+A^nB_2}. \tag{5.22}$$

The fractional gain changes caused by the fractional change, dA in A, are
obtained by differentiating these expressions to give

$$\frac{dK_1}{K_1} = \frac{n}{1+AB_1}\frac{dA}{A} \quad \text{and} \quad \frac{dK_2}{K_2} = \frac{n}{1+A^nB_2}\frac{dA}{A}.$$

But, because K_1 must equal K_2 for the two circuits to have the same gain,
$(1+AB_1)^n = 1+A^nB_2$ from (5.22), so

$$\frac{dK_2/K_2}{dK_1/K_1} = \frac{1}{(1+AB_1)^{n-1}}. \tag{5.23}$$

Let us examine what this means. If $n = 1$, the denominator of (5.23) is
unity so that the fractional gain variations are the same – an expected and
trivial result. But for values of n greater than unity and with $1+AB_1$, a
large positive quantity which is the usual case with considerable feedback,
then dK_2/K_2 will be less than dK_1/K_1.

Thus overall feedback would appear to be beneficial as far as stabilising
the gain is concerned. In practice, the number of stages that it can be
applied across is usually limited by stability considerations which are
described in the next section.

5.9 Instability

The forward gain of the amplifier, A, and the feedback fraction, B, are
properly represented by including both magnitude *and phase* at high and
low extremes of frequency. It is instructive to plot the product AB, the

181

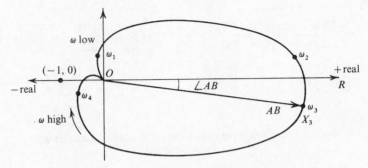

Fig. 5.13. Polar plot of AB, the loop gain, for a typical amplifier.
The Nyquist diagram.

loop gain, on a graph with polar co-ordinates. A typical plot for a multi-stage AC coupled amplifier is shown in fig. 5.13. At the frequency ω_3 the magnitude of the loop gain is the length of phasor OX_3 and its phase shift is the angle between OX_3 and the conventional positive real axis which represents zero phase shift. Instead of drawing a whole family of phasors similar to OX_3, one for each of the other frequencies, only the *locus* of X, the end of the phasor, is plotted with some frequencies shown on it.

At frequencies between ω_2 and ω_3, the phasor is nearly in the real direction. This means that the phase shifts are small and the magnitude of the loop gain is nearly a constant; a real, positive number. In an amplifier response plot, these will represent the 'midband' frequencies. The diagram also shows that there is rather more than 180° lagging phase shift at very high frequencies although the gain magnitude has dropped to a low figure. Similar phase shift, but leading, occurs at low frequencies for AC coupled amplifiers. In the case of a direct coupled amplifier which has no coupling capacitors between stages to give phase shift at low frequencies, the forward gain A will only have phase shifts at high frequency.

This type of diagram is called a Nyquist plot for AB, the loop gain. The possibility of an instability in the amplifier can be obtained by constructions on this diagram. It will be remembered from §5.2, case (*b*), at the start of this chapter, that if A or B changed sign so that $|1+AB| < 1$, then the gain with feedback, K, was greater than the gain of the circuit without feedback. This condition gives positive feedback and it means that an amplifier to which it applies will have a frequency response shaped like that in fig. 5.14. Instead of the gain with feedback, K being constant, it rises to a most unacceptable peak at high frequencies. This might be the sort of response obtained by putting overall feedback on a three-stage amplifier

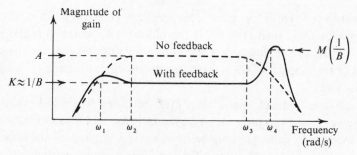

Fig. 5.14. Gain plotted against frequency for an amplifier with a tendency to instability at high frequency.

without careful design. How can we predict that this will happen? Alternatively, what sort of Nyquist diagram will be wanted for an amplifier that will give good, stable operation with a large AB factor?

†5.10   Interpretation of the Nyquist diagram

Fig. 5.15 shows a part of the Nyquist plot expanded in the region of the $(-1, 0)$ point, Y. At any frequency ω_4, taken at random, the magnitude of the loop gain is OX_4 and the phase angle, $\angle AB$ is as shown. Now remember

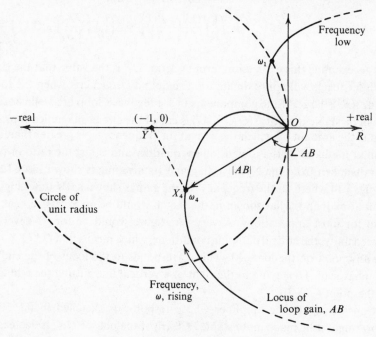

Fig. 5.15. Part of a Nyquist diagram.

that when $|1 + AB|$ is less than 1, the feedback becomes positive. If AB is the phasor OX_4, then $(1 + AB)$ is the phasor YX_4 where Y is the point $(-1, 0)$. This is obviously less than unity for any frequencies where the locus of the loop gain, AB, is inside the dotted circle of centre $(-1, 0)$ and radius unity.

A positive feedback condition is often tolerated to a small extent if it only gives a gain variation such as is shown in fig. 5.14 for low frequencies. The increase in gain *may* only be small but nevertheless all the beneficial effects of feedback such as stabilisation of gain, improved linearity, etc., are all reversed. Note that a single-stage amplifier with one capacitor coupling in the input circuit and one effective inertia (capacitances shunting the output) will only give 90° phase lead or lag, as shown in chapter 1. Thus the locus of the loop gain cannot enter the unit circle at any frequency and the feedback will always be negative.

In the Nyquist diagram of a particular amplifier, the phasor OX_4 might well lie inside the unity circle. To get an accurate figure for the value of the gain, we examine the expression for feedback and obtain values for its terms from the Nyquist plot. From (5.2),

$$\text{Gain with feedback} \quad K = \frac{A}{1+AB} = \frac{1}{B}\left(\frac{AB}{1+AB}\right)$$

$$= \left(\frac{1}{B}\right)\frac{OX_4 \angle ROX_4}{YX_4 \angle RYX_4} = \left(\frac{1}{B}\right)\left(\frac{OX_4}{YX_4}\right)\angle OX_4 Y.$$

$$(5.24)$$

Let us examine this expression term by term. $1/B$ is the value that the gain with feedback will approximate to at midband frequencies when the loop gain, AB, is large. If the components in the feedback loop are well chosen, it will be stable quantity. (OX_4/YX_4) is called the circuit magnification, M, and is the amount by which the gain at the frequency ω_4 is greater than the gain at midband. It is a dimensionless quantity and is just the ratio of two lengths taken from the Nyquist diagram. Its meaning is shown more fully on fig. 5.14 where, at the frequency ω_4, the gain is shown as M times higher than at midband. That shown in the sketch would be an excessive peak in gain for most applications. A very safe figure would be for M never to exceed unity, the critical damping conditions, which means that $OX/YX = 1$ for any point on the locus, X. This corresponds to the locus of the end of the phasor of AB staying to the right of a vertical line cutting the real axis at the point $(-\frac{1}{2}, 0)$.

Consider again the amplifier whose response is sketched in fig. 5.15. How might it be made more stable? Clearly if the phasor OX_4 is decreased in magnitude, then YX_4 will be larger and so the response will have a

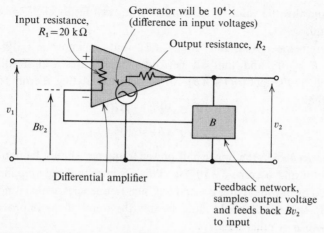

Input resistance, $R_1 = 20\,\text{k}\Omega$

Generator will be $10^4 \times$ (difference in input voltages)

Output resistance, R_2

v_1

Bv_2

Differential amplifier

B

v_2

Feedback network, samples output voltage and feeds back Bv_2 to input

Fig. 5.16. Amplifier and feedback arrangement for the worked example.

smaller peak in it. This would be achieved by reducing the forward gain A or the feedback B, so reducing $|AB|$. The other effects of such a simple basic method as this and the other methods available cannot be pursued in this introductory text. These topics are well treated in most books on control theory.

5.11 Worked example

An amplifier has a voltage gain $= 10^4$, an upper half power frequency 3×10^5 Hz and an input resistance $= 20\,\text{k}\Omega$. Voltage feedback is used to drop the gain to 100; what now is the upper half power frequency and the input resistance?

It is desired that, with feedback, the output resistance should be $10\,\Omega$ and the gain variation be 0.1 per cent for a 10 per cent supply voltage change: what are the maximum values of 'output resistance and gain variation which can be accepted in the amplifier before feedback is applied?

Solution. It is usually helpful to put the data given into some sort of diagram or figure. This is shown in fig. 5.16 and although no mention is made of the amplifier being a differential one, this will be the type most often encountered and so it is shown as such. The feedback is applied clearly to the inverting input. (The calculations will be identical for other voltage amplifier types such as that shown in fig. 5.2.)

The amplifier voltage gain of 10^4 is assumed to be the no-load value and the further assumption is made that the output load and feedback network draw only small currents so that the voltage drop across the output resistance is negligible. Now we can write

$$\text{Amplifier output} = v_2 = 10^4(v_1 - Bv_2) \tag{5.25}$$

185

and from this the value of the voltage gain with feedback, $K = v_2/v_1$, can be found.

But remember, that $K = A/(1 + AB)$, that we know $A = 10^4$ and we desire $K = 10^2$, and that all performance features with feedback are changed by the factor $(1 + AB)$. So without calculating B from (5.25), we can write

$$10^2 = 10^4/(1 + AB),$$

so

$$(1 + AB) = 10^2.$$

So with feedback, the upper half power frequency should be improved by this factor and become 3×10^7 Hz. This assumes that the amplifier gain is dropped by one main time constant at high frequencies, which is not always true. So this result should indicate only the trend of the improvement in the absence of other data.

The input resistance of a *voltage* amplifier when *voltage* feedback is applied will result in an improved voltage amplifier; namely one that draws *less* current from any source. So the input resistance will be 10^2 higher than before, and will be 2 MΩ.

With the other two features of performance, output resistance and gain sensitivity, we are told of the figures we desire after feedback and we know that these are 10^2 times better than those of the amplifier without feedback. Therefore the amplifier output resistance without feedback should be no more than $10 \, \Omega \times 10^2 = 1$ kΩ and the amplifier can be allowed to have a gain variation directly proportional to the supply voltage change.

Thus the performance figures for the amplifier without feedback are quite commonplace and yet those with feedback applied are good enough for most instrumentation purposes; namely, input resistance 2 MΩ, output resistance 10 Ω, half power frequency 30 MHz, and gain variation 0.1 per cent for expected supply changes. The trends shown by the simple feedback theory are accurate enough for most engineering applications – if more accurate results are wanted (particularly for frequency response) then the circuit would have to be tested or a more accurate model used for the amplifier. This is the subject of one of the problems at the end of this chapter.

5.12 Summary

Negative feedback is almost universally used in amplifiers purporting to have a good specification. It is economically favourable to use ordinary components with their accepted tolerances for most of the circuit elements and then to control the gain to a tight tolerance with a feedback circuit made of a few stable, expensive components. Negative feedback drops the gain of a circuit by the factor (1 + loop gain) but it improves almost all features of an amplifier performance by this same factor.

When feedback is put across a multistage amplifier, peaks in gain or even unwanted oscillations (instability) at the extremes of frequency may result. When an expensive system is being controlled, such as the attitude of a vertical take-off aeroplane, or the speed of a large machine, the system would first be tested 'open-loop' at various frequencies, and the loop gain would be measured in order to make a Nyquist plot. This allows any likelihood of instability to be predicted and corrected before the system is commissioned with the feedback connected. The possibility that the system runs into a wild, unstable state as a result of negative feedback becoming positive feedback is avoided.

5.13 Problems

1. Give three reasons for using negative feedback.

In fig. 5.17, the box represents an amplifier of gain -1000, input impedance 500 kΩ and negligible output impedance.

Calculate the voltage gain, and input impedance of the amplifier with feedback.

(Cambridge University: First year)

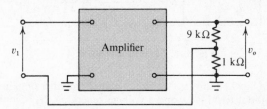

Fig. 5.17. Circuit for problem 1.

2. State the advantages derived from applying negative feedback to a high gain voltage amplifier.

The output circuit of an amplifier of gain A delivers a current I to a resistive load at a voltage V where V is related to the input voltage V_i by the equation

$$V = AV_i - RI.$$

Negative feedback is applied to the amplifier, a voltage $(\beta V + \gamma I)$ being fed back. Determine the output impedance of the amplifier after feedback has been applied, and the limiting value of this impedance as A becomes very large.

(Cambridge University: Second year)

3. A differential amplifier has a gain represented by:

$$v_3 = \left\{\frac{10^4}{1 + j\omega \cdot 10^{-5}}\right\} \times (v_1 - v_2),$$

187

where v_3 is the output voltage with no output load, v_1 and v_2 are the voltages at the two inputs 1 and 2 respectively.

The input resistance of the amplifier is 30 kΩ between inputs and the output resistance is 50 Ω. A feedback circuit is connected consisting of a 10 kΩ resistor from the output to input 2 and a 100 Ω resistor from input 2 to earth.

Determine the voltage gain of the amplifier between input 1 and the output at low frequency with feedback. What is the high frequency 'half power' point with and without feedback?

4. Develop and explain the general phasor diagram for an amplifier $A(j\omega)$ with negative feedback through a path $B(j\omega)$. Explain qualitatively, with reference to a phasor diagram, how negative feedback with $\angle B = 0°$ can reduce the effect of amplifier gain and phase variations on the overall transfer function.

An amplifier with gain A and zero phase shift is known to have an input resistance $r_{in} = 5$ kΩ. The amplifier has negative feedback applied through an ideal transformer with ratio B as shown in fig. 5.18 and a signal is applied through a resistance R_s. The overall gain V_o/V_s varies with R_s, and two sets of values are: $R_s = 25$ kΩ, $V_o/V_s = 5$; $R_s = 100$ kΩ, $V_o/V_s = 2$. Deduce the values of A and B.

(London University: Second year)

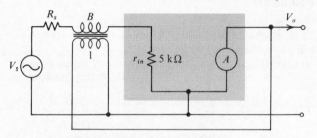

Fig. 5.18. Circuit for problem 4.

5. The circuit in fig. 5.19 shows a trans-impedance amplifier of gain R_4; current i flows from the input terminal A and a voltage generator $R_4 i$ is connected to the output terminal B.

When feedback resistors R_1 and R_3 are connected as shown, prove that the voltage gain is given by

$$\frac{v_2}{v_1} = -\frac{R_3}{R_1} \frac{1}{1 + \dfrac{R_3}{R_4}\left(1 + \dfrac{R_2}{R_3} + \dfrac{R_2}{R_1}\right)}.$$

If $R_1 = 1$ kΩ and $R_3 = 10$ kΩ, what possible values of R_2 and R_4 are required for the voltage gain of the circuit to be within 1 per cent of -10?

(Sheffield University, Second year)

6. Explain what is meant by *negative feedback* in an amplifier and discuss its effect on the performance of the amplifier.

Two identical voltage amplifiers have gains which may vary from 50 to 100. The gain is produced without phase shift.

If the amplifiers are to be used in series, with negative feedback of such magnitude that the overall gain does not fall below 100, determine the amount of feedback required and the maximum variation in gain, (*a*) when feedback is

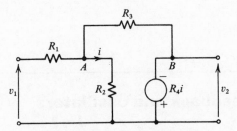

Fig. 5.19. Circuit for problem 5.

applied overall, and (*b*) when equal feedbacks are applied to each amplifier separately.

7. (*a*) Explain how negative feedback reduces the sensitivity of the gain of an amplifier to changes in the amplifier parameters.

An amplifier has a low frequency forward gain without feedback equal to −1000. When negative feedback is applied the closed-loop gain becomes equal to −200. What is the feedback factor, and what is the percentage change in closed-loop gain if the forward gain is reduced to −900?

(*b*) An amplifier has a gain–frequency response given by

$$A = -\frac{1000}{(1 + j\omega . 10^{-5})^3}.$$

What is the maximum value for the feedback factor if the amplifier is to be stable when negative feedback is applied?

(Birmingham University: Second year)

6

Positive feedback and oscillators

6.1 Introduction

The feedback discussed in chapter 5 was used to improve the performance of amplifiers and was broadly entitled 'negative feedback'. The later sections of the chapter explored the undesirable consequences for amplifiers of certain combinations of gain and phase shift round the feedback loop. However by designing a feedback circuit expressly to give phase shifts so that the feedback factor changes sign from negative to positive, a very useful series of switches, oscillators and signal generators can be made.

Every portable radio using the 'superhet' principle has a simple oscillator in it. Very highly stable oscillators are used as clocks and in precision test equipment. In all branches of engineering, the use of oscillators is very widespread and a variety of devices and techniques are used to make them but the same principles are used in all the applications. These principles of operation and design are explored in this chapter.

6.2 Theoretical requirements for oscillation

The overall gain, K, for an amplifier of gain A having feedback B applied from its output to input was derived in chapter 5 as

$$K = A/(1+AB). \tag{6.1}$$

Whenever the term AB is exactly equal to -1, then the gain K becomes infinite; that is, one can conceive of having an output for a negligible signal input and thus the circuit can be a generator. Described in this way, it may seem uncertain how one can make the circuit ever generate anything useful. However in the requirement for the circuit to be a generator, there are really two conditions stated in making $AB = -1$ which can also be written as $1\angle 180°$, and from this it may be clearly seen that the two conditions are:

(a) The magnitude of the gain round the loop shall be 1 to maintain oscillations (it needs to be greater than 1 to start them); and

190

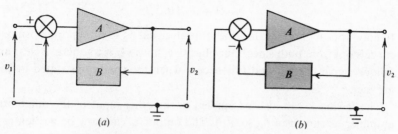

Fig. 6.1. Basic amplifiers and conditions for oscillation.
The circuit of (b) oscillates if $AB = 1\angle 180°$.

(b) The phase shift round the loop shall be 180° different from the condition normally required for negative feedback.

The latter requirement is made clearer if the basic amplifier block that was used for negative feedback is examined again. It was assumed that there was a differencing circuit in the path of the feedback signal (i.e. the difference of the input signal and the signal fed back was the amplifier input) as shown in fig. 6.1(a). Thus the circuit of fig. 6.1(b) will oscillate for the condition shown, namely, $AB = 1\angle 180°$. Fig. 6.2(a) now shows

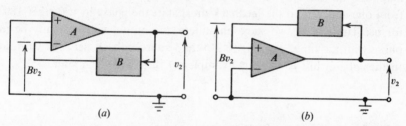

Fig. 6.2. Basic amplifiers and conditions for oscillation: (a) oscillates if $AB = 1\angle 180°$; (b) oscillates if $AB = 1\angle 0°$.

an oscillator using an integrated circuit differential amplifier. This circuit is identical to that of fig. 6.1(b). Since the ($-$) input to the amplifier is used, the feedback B will be required to have 180° phase shift at some frequency to give the oscillation condition. Then whatever the magnitude of B is at this frequency, if the gain, A, is large enough for $A \geqslant 1/B$, this gives a loop gain, AB, to be $\geqslant 1$, and the first condition for oscillation (a) is met also.

These conditions can also be obtained by using the ordinary circuit analysis for an amplifier. The output signal, v_2, can be written in terms of the inputs for the circuit of fig. 6.2(a) as,

$$v_2 = A(\text{difference of inputs}) = A(0 - Bv_2) \qquad (6.2)$$

191

or $$1 = -AB.$$

It may appear in (6.2) that the output signal v_2 is unimportant since it cancels out from both sides. Our objective, however is to obtain a clear and unique signal output, and this occurs when the conditions stated in §6.3 are met.

The feedback can also be applied to the other input of the differential amplifier as shown in fig. 6.2(*b*). The output v_2 can now be written as

$$v_2 = A(\text{difference of inputs}) = A(Bv_2 - 0) \qquad (6.3)$$

or $\quad +AB = 1 \quad$ which can also be written as $1 \angle 0°$.

Thus the oscillation requirement can be qualitatively described as follows: If a signal exists at the output such that after being fed back through B and amplified by A it arrives back exactly with the same amplitude, so $|AB| = 1$ and with the same phase, then it will be maintained and the circuit will be a generator.

The diagrams of figs. 6.2(*a*) and (*b*) can be generally applied to simple amplifiers made up of any number of stages. A single-stage or three-stage amplifier will normally have the input and output shifted in phase by 180° from one another, so the feedback must shift the phase by a further 180° for oscillation. For two-stage amplifiers, the output will normally be in phase with the input as in fig. 6.2(*b*) and so a feedback network with no further phase shift is wanted. Examples of these are given later.

6.3 Sinusoidal oscillators

In the design of a sinusoidal oscillator, two performance features need to be stable and well defined: (*a*) the frequency, and (*b*) the amplitude of the signal generated. These are usually independently but exactly linked to the two basic requirements for oscillations in the feedback amplifier for ease of design. Thus the most usual arrangement, shown basically in fig. 6.3(*a*) is as follows:

(*a*) At the desired frequency only, the phase shifts round the loop should be zero, so $\angle AB = 0°$. This condition may theoretically be achieved by any situation where the phase shift of the amplifier is exactly equal in magnitude to the phase shift of the B network but is opposite in sign.

However an amplifier with an odd number of stages, or a differential amplifier with an inverting input, has a phase change of almost exactly 180°. It is possible to have other phase shifts at very high and low fre-

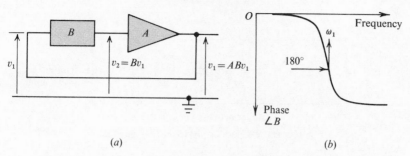

Fig. 6.3. (*a*) Oscillator arrangement, and (*b*) possible phase characteristic for network *B*.

quencies but $0°$ and $180°$ can usually be maintained stably over a wide frequency range for 'non-inverting' and 'inverting' amplifiers respectively. This was called the 'midband' frequency region in chapter 1.

To meet this first condition of oscillation, for an amplifier with a phase shift of, say, $180°$, the *B* network must provide a further phase shift of $180°$. Say that the *B* network has a phase–frequency relation of the kind shown in fig. 6.3(*b*). At a frequency ω_1, the condition for oscillation is met. Thus it is the *B network* and not the amplifier which sets the frequency of oscillation to be at or very close to ω_1.

(*b*) At the frequency ω_1, the magnitude of the loop gain, $|AB|$, should be greater than 1 for small signal levels and should become exactly unity for the desired output voltage amplitude. This can be made a feature of the amplifier only if some part of its circuit can be made *amplitude* conscious.

The two requirements to be met can be designed for separately. Because of the first requirement, namely, that the phase shift round the loop must be zero, such oscillators are called *phase-shift oscillators*. (In practice this covers *all* types of oscillators except those using the negative resistance characteristics of a circuit.)

6.4 Realisation of phase-shifting networks

Phase-shift networks may be constructed using resistors, capacitors, inductors and other devices. They are easier to make and cheaper when made of ordinary resistors and capacitors. Some examples of these are described first. Circuits using an inductor in a resonant circuit also have very clear frequency–phase relationships and are often used at radio frequencies (more than a few MHz). A specially sliced quartz crystal is a device which has an extremely sharp phase change with frequency in its

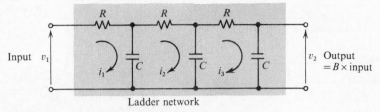

Ladder network

Fig. 6.4. An R–C ladder phase-shifting network.

resonant condition and so gives oscillators of good stability. The few examples described in the next sections show the principles of analysing all the types of phase-shift networks available.

6.4.1 Ladder networks

Consider the circuit shown in fig. 6.4. This is one type of R–C ladder network having three stages of R and C.

Let the network have an input voltage v_1; we desire to find an expression for the output voltage, v_2, so that we can find B which is the ratio v_2/v_1 or the 'gain' of the block. If currents i_1, i_2 and i_3 are the circulating currents in the loops of the networks, then Kirchhoff's voltage equations give:

$$v_1 = i_1 R + (i_1 - i_2) . 1/j\omega C,$$

$$0 = (i_2 - i_1)(1/j\omega C) + i_2 R + (i_2 - i_3)(1/j\omega C),$$

$$0 = (i_3 - i_2)(1/i\omega C) + i_3 R + i_3(1/j\omega C),$$

$$i_3 . 1/j\omega C = v_2.$$

These four equations contain five unknowns which are two voltages and three currents. Substituting to eliminate the three currents that are the unknowns in which we are not interested, we are left with one equation relating the two voltages. This is

$$\frac{v_2}{v_1} = \frac{1}{1 + 6j\omega CR + 5(j\omega CR)^2 + (j\omega CR)^3}$$

$$= \frac{1}{(1 - 5\omega^2 C^2 R^2) + j(6\omega CR - \omega^3 C^3 R^3)}. \tag{6.4}$$

The way that the relation v_2/v_1, which is B, is written in (6.4) shows clearly and separately the *real* part and *imaginary* part of the denominator. The phase shift will be $0°$ or $180°$ (we don't know which yet) when the imaginary term is zero; that is, when

$$\omega = 0 \quad \text{or} \quad \omega^2 C^2 R^2 = 6$$

$$\text{or} \quad \omega = (\sqrt{6})/CR. \tag{6.5}$$

The first solution, an oscillation at zero frequency, is hardly very useful but the other possibility gives a clear relationship by which to choose the product CR. Any suitable values of capacitance and resistance that give this product will be satisfactory with the proviso that resistors between 10^3 and 10^6 ohms and capacitors between 10^{-10} and 10^{-5} farads are usually widely available and cheap. Other values can of course be used but care must be taken to ensure when very low impedance values are chosen, that they do not load the amplifier unduly, or when very high impedance values are chosen, that they are not loaded by the amplifier. Also capacitors of more than $10\,\mu\text{F}$ (10^{-5} F) tend to be bulky and expensive if they are of types with good stability.

The value of frequency for which the phase shift is defined can now be substituted into (6.4) to give:

$$B = \frac{v_2}{v_1} = \frac{1}{1-5\times 6+\text{j}(0)} = -\frac{1}{29}. \qquad (6.6)$$

Since the condition for oscillation is that $AB \geqslant 1$, a suitable amplifier would be one whose gain magnitude was 29 or greater together with 180° phase shift between its output and input terminals.

Note that if the output resistance of the amplifier is appreciable, then its value can be taken into account by *reducing* the first resistor of the ladder by an amount equal to the output resistance. However this is not a good situation to have in oscillators of high stability as the output resistance of an amplifier depends on the components of which it is made and these may have considerable tolerances. Thus amplifiers of low output resistance are preferred; the methods of achieving this were dealt with in chapter 5, for example, by using local feedback.

More than three stages of R and C can be used in ladder networks. Also the capacitors and resistors can be exchanged in position. The method of analysis of these other circuits is similar to that just given.

6.4.2   Wien networks

The Wien network can be made in several ways, two of which are shown in fig. 6.5. The analysis of the circuits is similar to that described for the ladder networks, by putting in circulating currents or by treating the network as an attenuator of elements Z_1 and Z_2. Z_1 is R in series with C, and Z_2 is R in parallel with C in fig. 6.5(a). The equation relating v_2 and v_1 can be found as

$$v_2/v_1 = \frac{1}{3+\text{j}(\omega CR - 1/\omega CR)}. \qquad (6.7)$$

Now the imaginary part disappears when $\omega CR = 1$ which again allows

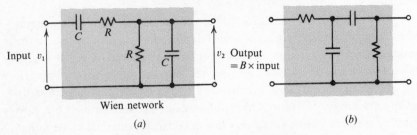

Wien network

(a)   (b)

Fig. 6.5. A Wien phase-shifting network.

the product CR to be selected to define the frequency of oscillation. Putting $\omega CR = 1$ into (6.7) gives

$$B = v_2/v_1 = \frac{1}{3+j(0)} = \tfrac{1}{3}\angle 0°. \tag{6.8}$$

Thus the amplifier required to put into a loop to make an oscillator with the Wien phase-shifting network is one with a gain $A \geqslant 3$ and no phase shift. Such an amplifier will be described in detail in §6.6.1 and in the worked example, §6.8.

The Wien network is very popular in oscillator circuits. Clearly, using only four components, it is very economical. With a gain requirement of only three times, a very simple single stage amplifier is suitable for use with it *or*, preferably, a higher gain amplifier may be used with a lot of negative feedback across it to reduce its gain to 3. The feedback will improve all the other performance features of the amplifier and this will lead to a better oscillator.

The other common R–C network used in oscillator circuits is the twin-T network. It contains more components and is harder to set up than the Wien network since the components must have well defined values relative to each other but it gives a very steep phase change with frequency and so will make a very stable oscillator (see §6.5).

†6.4.3 Tuned networks

Tuned circuits are widely used in frequency selective amplifiers, in radio sets and in instrumentation where some signal is to be emphasised or an undesired one filtered out. It is basically a circuit containing inductance, capacitance and resistance and its response can be analysed at various frequencies by writing equations for the circuit in the normal way. In particular, if it is to be used in the phase-shift network in an oscillator, we want an expression showing the magnitude and phase shift of the output/input voltage ratio.

Consider the circuit shown in fig. 6.6. The parallel tuned circuit consists

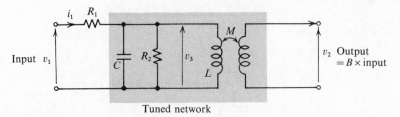

Fig. 6.6. A tuned (*LCR*) phase-shifting network.

of the inductor L and the capacitor C. R_2 is a resistance shunting the circuit which takes into account all the losses of the actual components. R_1 is the resistance of the source of voltage v_1. The output is shown taken from some secondary turns of a coupled circuit which has a mutual inductance M with the inductor L. This allows the circuit to be connected to the input of the amplifier when in an oscillator loop while keeping, quite separate, the steady voltages that may be developed either at the input or the output.

The equation for the voltages round the left-hand circuit is

$$v_1 = i_1 R_1 + v_3.$$

The equation showing that the current i_1 through R_1 must flow through the paths provided by C, R_2 and L respectively is

$$i_1 = v_3 j\omega C + (v_3/R_2) + (v_3/j\omega L).$$

The voltage output, v_2 will be, for no output current,

$$v_2 = j\omega M(\text{current through } L) = j\omega M v_3/j\omega L$$
$$= v_3 M/L.$$

From these three equations, two unwanted variables (i_1 and v_3) can be eliminated leaving one equation relating v_1 and v_2 which is

$$B = \frac{v_2}{v_1} = \frac{M}{L} \frac{1}{(1 + R_1/R_2) + j(R_1\omega C - R_1/\omega L)}. \qquad (6.9)$$

This equation will give $0°$ or $180°$ phase shift between v_2 and v_1 when the imaginary part disappears, which is when either,

$$R_1 = 0 \quad \text{or} \quad \omega C - 1/\omega L = 0,$$

so
$$\omega = 1/\sqrt{LC}. \qquad (6.10)$$

The first condition, that there will be no phase shift if R_1, the resistance of the voltage source v_1, is zero makes the network gain $B = M/L$ and therefore quite unselective with respect to frequency. Under these conditions,

197

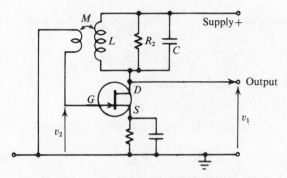

Fig. 6.7. A tuned oscillator using a f.e.t. amplifier.

the oscillator would give a square wave output when the amplifier gain is sufficient for oscillation.

The more useful case defines an angular frequency given by $1/\sqrt{LC}$ which substituted in (6.9) gives the gain of the network

$$B = (M/L)\{R_2/(R_1+R_2)\}.$$

An example of a tuned network in a simple oscillator is shown in fig. 6.7. The amplifier is a single-stage f.e.t. amplifier. In this simple circuit, the tuned circuit is the load in the drain circuit and the separate winding across which v_2 is developed is connected to the gate. This allows the bias between gate and source to be developed for the device by means of the decoupled resistor in the source lead as explained in chapter 2. Remembering that a single-stage amplifier has a 180° phase shift between input and output signal, B too will have to have a phase change of 180°. This is achieved by simply reversing the connections to the coil. (This changes the sign of M for those who want the algebra looking right.)

Other tuned circuits which could equally well be connected to the same type of amplifier are shown in fig. 6.8(a) and (b). They can be analysed in a similar way to that outlined here. The oscillators using them are called

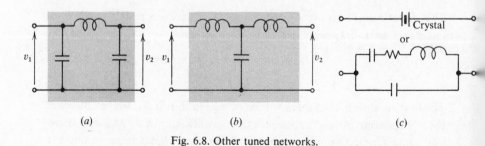

Fig. 6.8. Other tuned networks.

after their designers, Colpitts and Hartley for 6.8(a) and 6.8(b) respectively. Fig. 6.8(c) shows that the equivalent circuit of a quartz crystal is a tuned circuit and this device can also be used to make an oscillator. Readers wishing for details of these other oscillators should consult the more advanced texts listed in appendix A.

†6.5 Frequency stability

Any user of an oscillator will be interested in its frequency stability. In applications such as clocks, frequency generators, and precision timing circuits, where a high degree of frequency stability is imperative, special precautions have to be taken. The circuit construction must be very robust, the components carefully aged so that their value does not fluctuate. Often the components are mounted in a constant temperature oven to prevent any change in their values as the surrounding air changes temperature.

The need for these precautions is apparent when it is remembered that capacitors may have a temperature coefficient of 200 parts per million per °C. Thus a 0.2 per cent change in the frequency defined by the Wien network would result from a 10 °C temperature change which can easily be caused by draughts or intermittent sunlight.

However the oscillation requirement is that not only the phase shift of B, the feedback network, but that of the whole loop should sum to zero. When the amplifier has small but significant phase shifts due to stray capacities in its circuits or due to component value changes with age or temperature, or due to changes of the transit time of currents passing through the active devices, the frequency of the oscillator can drift. This drift is much reduced if the rate of phase change with frequency of the phase shifting network is very high. This is the slope of the characteristic in fig. 6.3(b) around ω_1. Then random, unwanted, phase shifts elsewhere in the circuit will make little difference to the frequency that the B network is defining. In other words, we are stating that $d\phi/d\omega$ should be high where ϕ is the phase shift given by the network in the region of the oscillating frequency ω_1.

The derivative $d\phi/d\omega$ can be calculated for any of the networks considered so far. To illustrate the method, the tuned network of §6.4.3 will be considered. Equation (6.9) for the circuit gave the phase change between v_2 and v_1 from the ratio of imaginary and real parts of the expression. So when ϕ is near 0° or 180°,

$$\phi = \tan \phi = \frac{-(R_1 \omega C - R_1/\omega L)}{1 + R_1/R_2} = \frac{-R_1 R_2(\omega C - 1/\omega L)}{R_1 + R_2}$$

199

and so
$$\frac{d\phi}{d\omega} = \left(\frac{-R_1 R_2}{R_1 + R_2}\right)\left(C + \frac{1}{\omega^2 L}\right).$$

Near the resonant frequency where $C = 1/\omega^2 L$ (by re-writing (6.10)),

$$\frac{d\phi}{d\omega} = -\frac{2}{\omega}\left(\frac{R_2 \| R_1}{\omega L}\right) = -\frac{2Q}{\omega}, \qquad (6.11)$$

where Q is the quality or Q-factor of the parallel tuned circuit. It is defined as the value of the resistive parallel paths divided by the impedance of the inductive parallel path at the frequency in question. (There are more general definitions of Q-factor.) It can be of the order of a few hundred with a carefully wound coil with a good core or, with a crystal, the equivalent Q can be of the order of thousands.

Equation (6.11) can be re-written to give $d\omega/\omega$ which is the fractional frequency change. Thus if a few degrees of variation in phase shift can be expected because of component defects in a given amplifier then, for a tuned circuit with a Q of 200, and $d\phi = 0.1$ radians,

$$-\frac{d\omega}{\omega} = \frac{d\phi}{2Q} = 2.5 \times 10^{-4} = 0.025 \text{ per cent}$$

or 250 parts per million. Note that with a crystal this frequency stability could be improved upon by an order of magnitude.

Note that the resistance, R_1, of the apparent source appears in (6.11) and that it should be large in order that $d\phi/d\omega$ is large enough to make the frequency stability good. This condition is met in both f.e.t. and bipolar transistors which have high output resistances in their models, namely, r_d or h_{oe} respectively. For best stability with tuned circuit networks, feedback might be employed to raise this apparent source resistance. Another way of looking at this is to make the amplifier feeding the tuned circuit into a good current source which, by definition, has a high shunting impedance.

To obtain the expression (6.11), C was eliminated and the expression was made into a relation about the quality of the inductor. The term L could have been eliminated instead and stability would then have been related to the R–C product or time constant of the circuit which is another measure of component quality.

6.6 Stabilisation of oscillator output amplitude

The second requirement to get oscillations from an amplifier and phase-shift network loop is that the magnitude of the loop gain, $|AB|$, must be greater than 1 for the oscillation to build up. The amplifier's gain, A, will

200

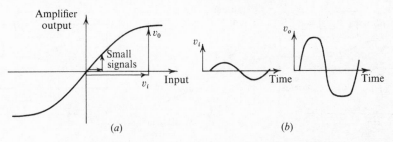

Fig. 6.9. Characteristics of an amplifier that is limiting.

be dependent on the tolerances of its components, the power supply voltage, the temperature, and any changes in the load connected to the output of the oscillator. So A will be liable to vary unless defined by feedback.

When $|AB|$ is much greater than 1, the signal amplitudes in the oscillator increase until there is 'limiting' at the amplifier output. Limiting effectively makes the amplifier gain lower when the signal amplitude is large and is shown in fig. 6.9(a). Here v_o/v_i, the gain, is clearly less when the signals are large. For small amplitudes, v_o varies almost linearly with v_i but for large values of v_i, the amplifier tends to saturate and the relation is non-linear. The output waveform is distorted as shown in fig. 6.9(b) and the circuit is not very satisfactory as a sinusoidal waveform source.

Feedback circuits can be used to reduce the gain of the amplifier when a defined signal output amplitude is reached. If this is less than the limiting level, then the amplifier input to output relation remains linear and so distortion does not take place.

Consider the amplifier shown in fig. 6.10(a) which has forward gain A and a feedback path so that $B_2 \times$ (output voltage) appears at the inverting input. This is an ordinary amplifier whose overall gain, K, between the other (non-inverting) input and output is $A/(1 + AB_2) \approx 1/B_2$ if AB_2 is large. If B_2 can be made to have a non-linear relation with output voltage v_2, then the desirable gain–output characteristic of fig. 6.10(b) is obtained. We already know that the conditions to make the system shown in fig. 6.10(c) into an oscillator are that $KB_1 =$ loop gain $= 1\angle0°$. At a frequency where the phase shift through the network B_1 is zero, the only remaining requirement is that the magnitude of the gain K is given by

$$|K| = 1/|B_1|.$$

Imagine now the oscillator being first switched on and consider the gain characteristic shown in fig. 6.10(b). There is no oscillation at first but, say

201

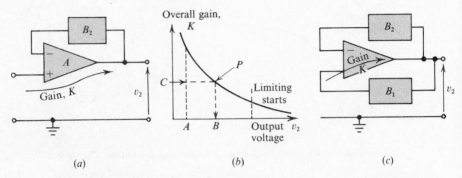

Fig. 6.10. Arrangement and characteristic of a variable gain amplifier and its connection as an oscillator.

as a result of device noise, the output voltage will be represented as being in the region A of the graph. This represents a high gain K and if this is greater than $1/B_1$, the oscillation will start to build up.

This means that the output voltage, v_2, rises to the region B where the intercept P on the graph gives a lower gain C. If this is equal to $1/B_1$, this is the output condition at which the oscillator will settle. This output voltage level should be well below the voltage at which limiting starts.

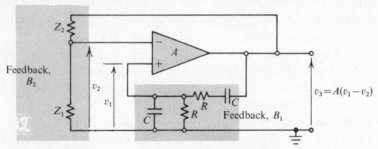

Fig. 6.11. A sine wave oscillator.

In fact feedback on the amplifier helps to improve its performance in other ways also. The simple oscillator now to be described gives such a pure sine wave that on fourteen specimens made for a laboratory experiment, the *worst* in performance gave only 0.1 per cent of unwanted second and third harmonic output voltage compared with the pure sine wave voltage. As well as working at voltage levels less than the limiting level, the amplifier linearity is very good due to the feedback. Also the output resistance is very low due to the voltage feedback.

The circuit shown in fig. 6.11 is exactly that shown in outline in fig. 6.10(c) except that B_1 and B_2 are shown as circuit elements. The network

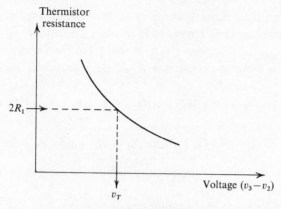

Fig. 6.12. Thermistor characteristic.

B_1 is a Wien phase-shift network which was shown in §6.4.2 to have no phase shift *and* a voltage output to input ratio, $B_1 = \frac{1}{3}$ at some frequency related to its component values. The gain K that we require is three, corresponding to the oscillation point P on fig. 6.10(b), the amplifier's gain characteristics.

The next section will explore a few of the ways of getting a gain characteristic of this sort.

6.6.1   Voltage sensitive amplifier using a thermistor

The overall gain of an amplifier, K, is defined closely by the components in its negative feedback network, B_2 which is made up of the impedances Z_1 and Z_2 as shown in fig. 6.11. Then

$$K = \frac{A}{1+AB_2} \approx \frac{1}{B_2} \quad \text{when} \quad AB_2 \gg 1$$

$$\approx \frac{Z_2+Z_1}{Z_1} \tag{6.12}$$

since $B_2 = Z_1/(Z_1+Z_2)$ if the effect of the source and load on the network B_2 can be ignored.

One possible way of making K decrease with output voltage is to make the impedance Z_2 decrease with voltage while keeping Z_1 constant (K and Z_2 appear in the numerator on each side of (6.12)). A suitable component for Z_2 is a thermistor whose characteristic is shown in fig. 6.12. The resistance of a thermistor falls as its temperature rises, as a result of an increase in the applied voltage. Special thermistors with very small beads of semiconductor material deposited between two fine wires in a sealed glass envelope are available for these applications. With a few milliwatts of

203

applied power, their temperature can rise by at least ten times the normal small changes of ambient temperature in which the circuit may operate.

Since a range of thermistors is available from the various makers, we wish to know how to choose one to define a particular oscillator output, v_3.

The voltage, v_T across the thermistor is given in terms of v_3 from fig. 6.11 as

$$v_T = v_3 - v_2 = \tfrac{2}{3}v_3 \quad \text{when} \quad Z_1 = R_1 \quad \text{and} \quad Z_2 = 2R_1.$$

From (6.12), this ratio of impedances defines the current gain K for the Wien network.

Therefore the thermistor to choose will be one where $\tfrac{2}{3}$ of the desired oscillator output voltage will be developing between a $\tfrac{1}{4}$ and $\tfrac{1}{2}$ of the maximum allowable dissipation into the thermistor. Then ambient temperature changes will have only a small effect. Lastly R_1 is chosen to be half the calculated thermistor resistance (for the Wien oscillator).

The only consideration, not mentioned so far, in the choice of the thermistor and the resistance R_1 is the load placed on the amplifier by these components once oscillations start. The total load is $Z_1 + Z_2 = 3R_1$. This load should be of the order of a few kilohms with most integrated circuit amplifiers to be quite sure of avoiding distortion due to overloading the integrated circuit.

When the oscillator is first switched on, the thermistor is cold and has a high resistance so the oscillation builds up rapidly. The rate of settling depends on the thermistor's own time constant so an overshoot or 'bounce' is usually noticed.

Lastly, when needed, a large output voltage can be developed in this oscillator without increased distortion. This is because the amplifier linearity is good as a result of the heavy negative feedback.

†6.6.2 Voltage sensitive amplifier using a bulb or a field-effect transistor

Equation (6.12) relates the component values in the negative feedback network to the value of B_2 and hence to the gain of the amplifier. In the last section, having Z_2 variable was but one method of design. An alternative is to make Z_2 a constant resistor, R_2 and have Z_1 as a component with a positive coefficient of resistance with voltage. Such a device is a small light bulb, or a positive temperature-coefficient resistor which has a characteristic like that sketched in fig. 6.13. Although a bulb has the advantages of cheapness and simplicity, it usually requires a considerable power input

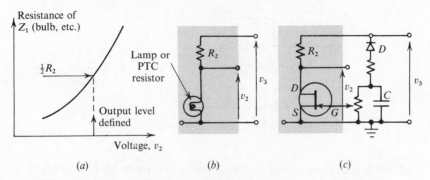

Fig. 6.13. Circuits with positive coefficients of resistance for stabilising an oscillator output amplitude.

to heat it. Both devices are basically temperature sensitive so the oscillator output amplitude will to some degree depend inversely upon the ambient temperature.

An alternative circuit using a f.e.t. is shown in fig. 6.13(c). For small values of drain-source voltage, v_2, which are below the pinch-off shown in the f.e.t. characteristic, fig. 2.7, the field-effect transistor behaves like a resistor whose magnitude depends on the value of the bias applied to the gate. It becomes, in effect, a voltage controlled resistor. With no applied gate voltage, the channel is fully conducting and so is a low resistance path. This low resistance, with R_2 fixed, will allow only a little of the signal v_3 to be fed back and so the amplifier gain which is the inverse of this fraction will be high and oscillations will build up. When this happens, the oscillator signal v_3 is rectified by the diode D and developed as a steady negative voltage across C. This is the correct polarity of bias to stop the transistor channel conducting as much as with no signal and its resistance increases. The variable resistor across C allows the circuit to be adjusted to give various output amplitudes but these should not be so big as to swing the f.e.t. (which receives $\frac{1}{3}v_3$) beyond pinch-off because the output waveform will then become non-sinusoidal.

The examples described here have used integrated circuit differential amplifiers. The same principles are used with amplifiers made of discrete components. To obtain very low frequencies (say 0.001 Hz) and very high frequencies (10 MHz and above), the reader will have to consult a more advanced text, but the principles used will be identical to the ones described here.

205

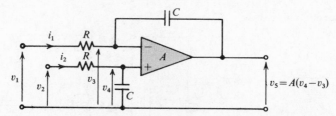

Fig. 6.14. Differential integrator.

†6.7 Oscillator using operational amplifiers as integrators

An interesting oscillator is provided by two integrators (see §4.7.3) in a loop. Because the concept of stage gain and phase angle is not usually developed for integrators, we consider first the circuit shown in fig. 6.14. If the amplifier has a no-load gain of A and if the currents drawn by C and any following circuits give negligible voltage drop across the output resistance of the amplifier, we can write

$$v_5 = A(v_4 - v_3).$$

If we consider the input resistance of the amplifier to be very high, then the input currents from v_1 and v_2 will flow on into the capacitors. Expressions for these currents give, respectively,

$$i_1 = (v_1 - v_3)/R = (v_3 - v_5)j\omega C,$$

$$i_2 = (v_2 - v_4)/R = v_4 j\omega C.$$

From these three equations, we can eliminate two unwanted voltages, v_3 and v_4, to write the output, v_5, in terms of the inputs, v_1 and v_2, as

$$v_5 = (v_2 - v_1)/(j\omega CR + (1 + j\omega CR)/A). \tag{6.13}$$

Now if A is large we can rewrite this as

$$v_5 = (v_2 - v_1)(-j/\omega CR)$$

$$= (v_2 - v_1)\left|\frac{1}{\omega CR}\right| \angle -90°. \tag{6.14}$$

Thus the circuit appears to have a gain magnitude of $1/\omega CR$ and a 90° phase lag for the input v_2 and a 90° phase lead for the input v_1 because $-v_1$ appears in (6.14).

Now consider the circuit shown in fig. 6.15. The amplifier A_1 with no input to the inverting terminal has a phase lag of 90° while A_2 has a phase lead of 90° (from (6.14)) and so the phase-shift relation for an oscillator

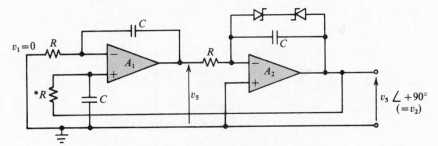

Fig. 6.15. Sinusoidal oscillator with quadrature outputs.

loop is obeyed. The gain relation for the loop is that each stage has a gain $1/\omega CR$ and so the gain round the loop is unity when

$$\omega = 1/CR \quad \text{or} \quad \text{frequency} = 1/2\pi CR \text{ Hz.} \tag{6.15}$$

One can analyse the circuit to show that slight mismatching of the components will cause the oscillation amplitude either to rise slowly or to decay slowly. In practice the resistor marked with an asterisk in fig. 6.15 is made slightly larger than the other resistors to help oscillations to build up. The Zener diodes across the second integrator conduct when the output amplitude is large and so limit it.

The feature of this oscillator is that although it uses two integrated circuit amplifiers, it gives useful outputs from the two amplifiers. If the output v_5 from A_1 is considered as a sine wave, then the output from A_2, which leads it by 90°, is a cosine wave.

6.8 Worked example

A two-stage f.e.t. oscillator uses the phase-shifting network shown in fig. 6.16. If the input resistance of the amplifier is very high, prove that

$$v_2/v_1 = 1/\{3 + \mathrm{j}(\omega RC - 1/\omega RC)\}.$$

Show that the frequency of oscillation is $f = 1/2\pi RC$ and that the gain of the amplifier, A must exceed 3.

If the amplifier has an input resistance = 1 MΩ and output resistance = 10 kΩ, and the capacitors of the network are 0.01 μF, what resistors are wanted for an oscillation frequency of 160 Hz?

Solution. A relation for v_1 and v_2 at the input and output of the phase-shift network is derived first. The effect of the amplifier on the network will be considered afterwards. For the circulating currents shown, Kirchhoff's voltage law gives:

$$v_1 = i_1 R + (i_1 - i_2) \cdot 1/\mathrm{j}\omega C, \tag{6.16}$$

$$v_1 = i_1 R + i_2(R + 1/\mathrm{j}\omega C), \tag{6.17}$$

$$v_2 = i_2 R. \tag{6.18}$$

207

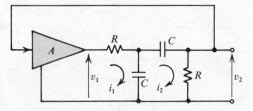

Fig. 6.16. Oscillator circuit for the worked example.

Subtracting (6.17) from (6.16) gives

$$i_1 = i_2 \left(\frac{1}{j\omega C} + \frac{1}{j\omega C} + R \right) j\omega C = \frac{v_2}{R}(2 + j\omega CR).$$

Hence (6.17) can be written as

$$v_1 = \frac{v_2}{R}(2 + j\omega CR)R + \frac{v_2}{R}\left(\frac{1}{j\omega C} + R \right)$$

$$= v_2\{3 + j(\omega CR - 1/\omega CR)\}. \qquad \text{Q.E.D.}$$

For a normal amplifier with $0°$ or $180°$ phase shift, the feedback network must also have $0°$ or $180°$ phase shift for oscillation to be possible. This requires the imaginary part to be 0 in the equation relating v_1 and v_2, so

$$\omega CR - 1/\omega CR = 0,$$

i.e. $$f = \frac{\omega}{2\pi} = \frac{1}{2\pi RC}.$$

With no imaginary part in the equation that was proved, the relation for v_1 and v_2 becomes,
$$v_2/v_1 = \tfrac{1}{3}.$$

So the required amplifier must have a voltage gain $A \geqslant 3$ for the loop gain $\geqslant 1\angle 0°$.

If we are to make the oscillator work at 160 Hz with capacitors, $C = 10^{-8}$ F, $$f = 160 = 1/2\pi R . 10^{-8},$$

therefore $$R \approx 10^5 \, \Omega.$$

Now the amplifier output resistance is 10 kΩ, and this is in series with the left-hand resistor in fig. 6.16. The total resistance is to be 100 kΩ, so the component needed for the circuit resistor is 90 kΩ.

Also the amplifier input resistance is 1 MΩ and this shunts the right-hand resistor in fig. 6.16. If the total resistance is to be 100 kΩ, the component needed is about 110 kΩ.

6.9 Summary

Inadvertant 'positive feedback' that gives rise to unwanted oscillations can be disastrous to equipment performance and troublesome to cure. However when deliberately used, and carefully controlled, it can give sine and square wave generators of great stability.

For ease of design, it is better to separate the parts of the circuit that control the two basic conditions for oscillation; namely that the phase shift round the loop is zero and the magnitude of the loop gain is $\geqslant 1$. By making a phase-shift network with a high rate of change of phase with frequency, the small, unwanted but variable phase shifts in the amplifier can be made to have little effect. The actual phase shift round the loop then largely depends on a few components in the network which can be selected for their high stability.

The gain of the amplifier must offset the attenuation in the phase-shift circuit so that the loop gain is greater than 1 to start oscillation. The loop gain should then fall to 1 at some output amplitude below the limiting level of the amplifier output. This gives a relatively undistorted sinusoidal output. Some ways of achieving this 'automatic gain control' have been described.

Throughout the chapter, the general principles of feedback have been described and no attempt has been made to catalogue the variety of circuit designs that can be constructed. Integrated circuits can be used to make amplifiers with great success and this is given prominence in the chapter.

6.10 Problems

1. Fig. 6.17 shows the schematic circuit for a phase-shift oscillator. The voltage amplifier, A, may be assumed to have a very high resistive input impedance, the effect of which on the rest of the circuit is negligible. The amplifier introduces no phase shift.

Obtain an expression for the frequency at which stable oscillation may occur and show that the overall voltage gain of the amplifier must be at least 3 for this to be so.

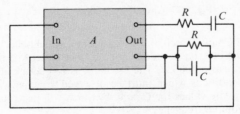

Fig. 6.17. Circuit for problem 1.

209

Determine the frequency of oscillation when $R = 10\,\text{k}\Omega$ and $C = 0.005\,\mu\text{F}$. Explain why this type of oscillator is mostly used for producing oscillations at audio frequencies whilst an oscillator incorporating a tuned resonant circuit is more suitable for radio frequencies.

(Cambridge University: Second year)

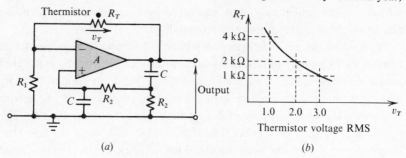

(a)   (b)

Fig. 6.18. Circuit for problem 2.

2. An oscillator arrangement using an integrated differential amplifier is shown in fig. 6.18(a). The characteristic of the thermistor in the circuit is sketched in fig. 6.18(b).

Suggest values for R_1, R_2 and C for a 10 Hz sine wave oscillator of 3 V RMS amplitude output. What assumptions are made to solve the problem with the given data?

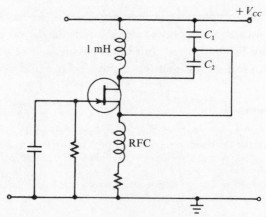

Fig. 6.19. Circuit for problem 3.

3. Fig. 6.19 shows a Colpitt's oscillator using a junction field-effect transistor as the active device. The f.e.t. has a mutual conductance equal to 5 mA/V and an output impedance equal to 100 kΩ. The oscillator is to operate at $10/2\pi$ MHz. What are the two conditions which must be observed to achieve oscillation at this frequency? Calculate values for C_1 and C_2 which will just cause the circuit to oscillate at the required frequency.

(Birmingham University: Third year)

Problems

4. Explain the importance of the Nyquist criterion of stability in the study of feedback amplifiers and show how the stability of such an amplifier may be predicted from a knowledge of its open-loop amplitude–frequency and phase–frequency responses.

An R–C oscillator consists of a phase-reversing amplifier of gain m and a phase-shifting network, connected as a closed loop. The transfer function for the network is $\phi(j\omega)$ where:

$$\phi(j\omega) = [1 + j\omega T]^{-1} [1 + 2j\omega T]^{-2}.$$

Find the minimum value for m for which the system oscillates and the frequency of oscillation.

(Newcastle University: Third year)

211

Quiz 3 (chapters 5 and 6)

Underline the correct statements.

1. The effect of negative feedback on an amplifier is to (a) increase gain, (b) decrease gain, (c) make the amplifier oscillate, (d) improve stability.

2. A feedback amplifier with $A = 1000$ and $B = 0.20$ will have an overall gain of about (a) 20, (b) 200, (c) 50, (d) 5, (e) -5.

3. The amplifier of question 2 is a voltage amplifier and has 1 V of unwanted hum at its output before voltage feedback is applied. After feedback the hum is (a) 1 V, (b) 0.2 V, (c) 5 mV, (d) 1 mV.

4. The amplifier of question 3 has an input resistance of 1 kΩ before voltage feedback is applied. After feedback the input resistance is (a) 200 kΩ, (b) 5 Ω, (c) 1 MΩ, (d) 5 kΩ.

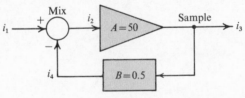

Fig. 1

5. The amplifier of question 2 is a current amplifier and has an output resistance of 100 Ω before current feedback is applied. After feedback its output resistance is (a) 0.5 Ω, (b) 20 kΩ, (c) 500 Ω, (d) 20 Ω.

6. Given i_2 in fig. 1, i_1 is (a) $26i_2$, (b) $101i_2$, (c) $24i_2$, (d) $99i_2$, (e) $6i_2$.

7. The overall current gain, i_3/i_1, of the circuit in fig. 1 is (a) just above 2, (b) 25, (c) just below 2.

8. Instability in the negative feedback amplifier of fig. 1 can be caused by (a) large phase-shifts at high frequency, (b) large phase shifts at low frequency, (c) loop gain tending to zero, (d) loop gain, AB, tending to -1.

9. The condition for continuous oscillation for the loop shown in fig. 2 is (a) $AB \geqslant 1\angle 180°$, (b) $AB = 2\angle 0°$, (c) $AB \geqslant 0.9\angle 0°$, (d) $AB \geqslant 1\angle 360°$.

212

Quiz 3

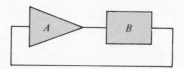

Fig. 2

10. An oscillator is made of an amplifier of zero phase shift and adjustable gain A and a network B where $B = 1/(5+\omega) + j(4-\omega^2)$. The oscillation frequency is given by (a) $\omega = 5$, (b) $\omega = 4$, (c) $\omega = 2$.

11. The circuit of question 10 will oscillate when A is (a) 5, (b) 7, (c) 1/7, (d) 3, (e) 9, (f) -7.

12. To make an oscillator with a stable output frequency, the desirable features are (a) high loop gain, (b) phase-shift network of low input and output resistances, (c) amplifier working well within its half power frequencies, (d) a phase-shift network with a high rate of change of gain as frequency is varied, (e) a phase-shift network with a high rate of change of phase as frequency is varied.

13. To make an oscillator with a stable output amplitude, the desirable features are (a) a very stable phase-shift network, (b) an amplifier with a high rate of change of gain with output voltage amplitude, (c) an amplifier working well within its half power frequencies.

7

Digital fundamentals

7.1 Introduction

In chapters 2, 3 and 4, the capabilities of devices and integrated circuits have been explored for inputs which are continuous functions of time and where a *linear* output/input or 'gain' relation is usually sought.

Digital signals, where only two levels are considered, HIGH or LOW, have the features introduced in § 1.10 and can be realised with gates. At the expense of some added complexity in circuitry, a system, such as a voltmeter or communication channel, and many other such pieces of equipment, can be built with a great capability for accuracy or for noise rejection compared to analogue methods. The basic circuit of a gate is both simple and cheap. From a description that, for some readers, should relate to school experiences with multivibrators, simple clocks or Venn diagrams, the methods of analysing and designing simple digital circuits are developed in this chapter.

7.2 Transistor as a switch – the simple inverter

The digital mode of operating a transistor is simpler than the analogue one described in earlier chapters. Figs. 7.1 (*a*) and 7.1 (*b*) show a MOS and a bipolar transistor circuit respectively in which a change of input voltage controls the current put into the load, which may be a lamp, a bell, or a relay coil which needs to be energised. It is an important step to realise that, in digital circuits, the state *between* on and off is considered to occur *only briefly* between the normal conditions of fully ON or OFF. Thus a lamp glowing dimly or a relay which is just hovering on its energising condition is clearly avoided by a designer.

Consider the device output characteristics which are shown in fig. 7.1 (*c*). Here the output voltage is plotted against current and in chapter 2 the idea of a load line across this characteristic, to define the possible current and voltage states which could occur simultaneously, was used many times.

214

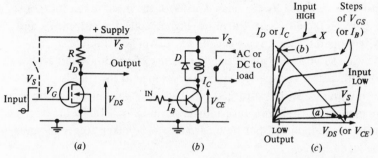

Fig. 7.1. Inverter circuits: (a) MOS, (b) bipolar, (c) typical device characteristic.

The position where the load line and the device characteristics intersect defines the circuit operating point. Thus with the *input* LOW, the intersection is at (*a*) in fig. 7.1(*c*) which corresponds to the device *output* voltage, V_{DS}, being HIGH and close to the supply V_S: thus a HIGH output is given by a LOW input.

Consider next the input voltage being HIGH which defines the device output voltage against current relation being, say, *X* on fig. 7.1(*c*). This intersects the load line at (*b*), which has a LOW value of output voltage V_{DS}: thus a LOW output is given by a HIGH input.

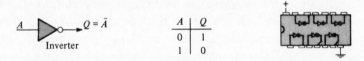

Fig. 7.2. The inverter: (a) symbol, (b) truth table, (c) typical package.

The circuit satisfies the two conditions needed for an inverter whose symbol is shown in fig. 7.2(*a*): the circle on the output line denotes inversion. The Boolean convention for NOT is a bar over the symbol so, for the signals shown, we write $Q = $ 'NOT A' $ = \bar{A}$. The inverter's truth table is in fig. 7.2(*b*): here LOW is shortened to 0 and HIGH to 1; the latter is taken to mean somewhere close to the supply voltage of the circuit. In § 7.10, the actual circuit details will be described.

For most logic applications, the inverter is bought in a 14-pin package of six identical gates, whose outline is in fig. 7.2(*c*). The cost is about that of a small ice-cream cone (it may interest readers in the future to do this comparison again). For some purposes where isolation is needed, say to control the current drawn by a load from an AC source, the circuit of fig. 7.1(*b*) is used. The circuit clearly has similarities to fig. 7.1(*a*) but the

215

resistor R is replaced by the relay coil. With the input taken HIGH, a small base current I_B will control a much higher collector current I_C and the relay will operate: the circuit is called a 'driver' or 'buffer'.

Mention should be made of the importance of the diode across the relay coil. It prevents a voltage spike being developed at the transistor collector each time the transistor is turned off. This may give an avalanche breakdown in the transistor, and uncertain device life. Some relays are now made with the diode included but then care has to be taken to get the polarity of the leads correct.

7.3 Simple gates – NAND, AND, NOR, OR, Exclusive-OR and NOR

It may help some readers to see the idealised circuit of a simple logic gate to show how little more complex it is than the inverter. A little consideration will show that in fig. 7.3(*a*) it is only when *both* the inputs A AND B are HIGH, that the output is changed as the two transistors are in *series*. The circuit can then draw current from the supply or discharge a capacitor C_o at the output so defining a LOW output. This situation is shown in the bottom line of the truth table fig. 7.3(*b*). With either input A LOW, B LOW or both LOW, there can be no complete path through the transistors, and so the output voltage (or any load C_o) will be charged up to the supply voltage and will be HIGH – a situation shown in the top three lines of the truth table. If the truth table output is inverted only one combination of inputs gives a HIGH output and we have the AND operation; namely, A AND B HIGH gives output HIGH. So an 'invert + and' or NAND gate was first described and its symbol is shown in fig. 7.3(*c*): the small circle on the output denotes the invert process and is similar to that in fig. 7.2(*a*).

The Boolean symbol for AND is a dot ($\cdot$) (in some schools and mathematical texts, $\cap$ or $\wedge$ is used); sometimes the dot is omitted. Using also the NOT or bar symbol, we can write for the NAND gate with inputs A, B, C, etc.

$$\text{Output} = (A \text{ AND } B \text{ AND } ...) \text{ inverted } = \overline{A \cdot B \cdot C} ...$$

or sometimes written $\overline{ABC} ...$

The symbol for an AND gate is shown in fig. 7.3(*d*); packages with two, three, four or eight inputs can be purchased.

The schematic circuits shown in figs. 7.1 and 7.3 have other components added to make their outputs follow any input change with little delay even

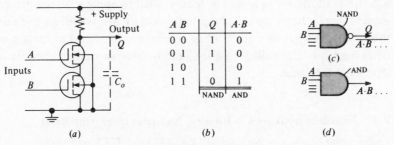

Fig. 7.3. (*a*) Simplified circuit of 2-input gate, (*b*) truth table of NAND and AND gates, (*c*) NAND symbol, (*d*) AND symbol.

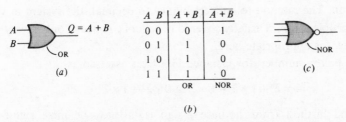

Fig. 7.4. (*a*) OR gate symbol, (*b*) truth table of OR and NOR gates, (*c*) NOR symbol.

if the output lines are loaded with a considerable capacitance. This delay is called 'propagation delay'. The full circuits are described in § 7.10.

The OR gate has an output which is HIGH if either one or the other or both of its inputs are HIGH. The symbol for the gate is shown in fig. 7.4(*a*), its truth table is in fig. 7.4(*b*), and the Boolean symbol for OR is + (in some texts ∪ or ∨). So $Q = $ 'A OR B' is written $A + B$. The truth table shows also the output state for a NOR gate which is just an inversion of the OR state: the symbol for this gate has the usual small circle added and is shown in fig. 7.4(*c*)

The last of the simple gates is the Exclusive-OR whose output is HIGH if either one or the other input is HIGH (but not both). It never has more than two inputs. Its symbol is shown in fig. 7.5(*a*), the truth table is in fig. 7.5(*b*), the Boolean symbol for the function is ⊕ and the name Exclusive-OR is

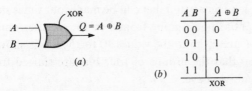

Fig. 7.5. The Exclusive-OR gate: (*a*) symbol, (*b*) truth table.

217

sometimes shortened to XOR. The reader is left to apply a similar arrangement to that in the paragraph above to think of the symbol for an Exclusive-NOR gate, and what its truth table is. Its name is shortened to XNOR sometimes, and it is also called a parity gate. It gives an output when its two inputs are the same.

7.4   Number systems – binary, hexadecimal and BCD

Engineering quantities such as 313 kg, 473 m/s, 440 V are assumed to be in decimal notation. In a computer or digital instrument, we need to represent these by HIGH or LOW states only, which is a 2-level or binary notation. The method to convert binary to decimal, the system in which most people think, is usually covered in school courses so only a very brief treatment is appropriate here.

The binary number, for instance 11001_2, is assessed as

$$1 \times 2^4 + 1 \times 2^3 + 0 \times 2^2 + 0 \times 2^1 + 1 \times 2^0 = 25_{10}.$$

The end furthest from the base 2 sign is taken to be most significant. This is exactly the same as our understanding of twenty-five as

$$2 \times 10^1 + 5 \times 10^0.$$

The converse or changing a decimal to a binary number is achieved by dividing the number successively by two and seeing if there is a remainder. Thus to convert 25_{10} to binary,

$$25 \div 2 = 12 \text{ remainder } 1$$
$$12 \div 2 = 6 \text{ remainder } 0$$
$$6 \div 2 = 3 \text{ remainder } 0$$
$$3 \div 2 = 1 \text{ remainder } 1$$
$$1 \div 2 = 0 \text{ remainder } 1$$

from which we read off the binary number as the remainder column, starting from the bottom, as 11001_2.

The reason for using the hexadecimal or hex system (base 16) is to shorten the binary representation. The numbers 0 to 15_{10} can be presented as four bits so a long binary number can be made four times shorter in hex. However, we want symbols going from 0 to 15; so, after numbers 0 to 9, the letters A to F are given to the values 10 to 15; this is shown in fig. 7.6. With the proviso that the groups of four bits are started from the least significant bit and, we have,

$$705_{10} = 1011000001 = 10\ 1100\ 0001 = 2C1_{16}$$

Number systems

Decimal	Binary	Hexadecimal
0	0 0 0 0	0
1	0 0 0 1	1
2	0 0 1 0	2
3	0 0 1 1	3
⋮	⋮	⋮
9	1 0 0 1	9
10	1 0 1 0	A
11	1 0 1 1	B
12	1 1 0 0	C
13	1 1 0 1	D
14	1 1 1 0	E
15	1 1 1 1	F
	MS LS	

MS = most significant
LS = least significant

Fig. 7.6. Comparison of binary and hexadecimal codes.

Conversion in the opposite direction is achieved as

$$2Cl_{16} = 2 \times 16^2 + 12 \times 16^1 + 1 \times 16^0 = 2 \times 256 + 12 \times 16 + 1 = 705_{10}.$$

Wide use of hexadecimal numbers arises from the commonly available 8-bit, 16-bit and 32-bit computers; in each case, the number of bits are divisible by four. An 8-bit code is used universally to represent the letters, numbers and symbols on a typewriter keyboard using the American Standard Code for Information Interchange (ASCII). So a group of eight simple 1-bit memory elements will store a typewriter symbol, or perhaps a simple microcomputer instruction, or a number (as part of a calculation) in the range $0-255_{10}$.

Such a computer may have eight lines (or a bus) leading from the processor chip to a memory chip or to an output socket. The state of these eight lines was indicated by lights in the past but is now a hexadecimal or decimal display. Thus if the lines were in the state 11000001, the display would probably be on two 7-segment indicators as Cl (or in decimal as $12 \times 16 + 1 = 193$).

Many digital instruments such as timers and voltmeters use the binary coded decimal or BCD representation as the display in such instruments is needed in decimal form. So a 4-bit group is made to represent *each* displayed decimal digit from 0000 for a zero to 1001 for a nine. To represent 139_{10}, 12 bits are wanted as

$$139_{10} = 0001\,0011\,1001\,(BCD).$$

The code is wasteful as many states are not used but decimal displays have important applications. Devices with 4 input lines named 'BCD decoder,

219

latch and display' come in the form of chips with a transparent top which lights up with a different decimal digit for each 4-bit code of high and low levels on its input wires.

The way of treating whole numbers (integers) as opposed to decimals (floating point numbers) and the way of achieving economy of computing with positive and negative quantities is left to microcomputer texts.

7.5 Definitions of combinational and sequential logic

In digital circuits, such as comparators, two 8-bit numbers may be taken to generate an output only when they are the same. Such circuits and others such as adders, multipliers, can all be built from the basic NAND and NOR gates: they are called *combinational* logic circuits because the output is defined by the inputs at any one instant and not by any memory of what these levels have been. These circuits can all be analysed and designed using the algebra proposed by George Boole, an Irishman, as long ago as 1854.

A separate requirement in electronics is to have a circuit whose next output depends on memory of the past 'sequence' of inputs and outputs. An example is a circuit which could sum up the number of 1's in a sequence or would generate the outputs in the right order needed for a street crossing traffic light. Thus the moment after all four red lights are operated, the circuit would have to remember which roads last had the green lights and give other roads the 'go' next. The basic element in all *sequential* logic is to have a simple *memory* element. We see in § 7.9 how the element can be made by *feedback* across two simple inverting gates, known to many as 'flip-flops'.

However, we will start with gates and combination logic since gates are basic to all digital circuits.

7.6 Combinational logic identities – Boolean algebra

In § 7.3, we have already met a number of simple gates and have seen that an AND gate with inputs A, B, C would have its output written as $A \cdot B \cdot C$ (or ABC) and an OR gate with similar inputs would have its output written as $A + B + C$. Fig. 7.7 shows a table of logic identities which are in two columns broadly showing the relations for variables combined using AND on the left and using OR on the right. Many of these are obvious from our knowledge of other algebras. However, Boolean algebra is one which relates variables which can *only* be in one of two states, e.g. HIGH or LOW

Combinational logic identities

		AND	OR
Associative	1(a)	$A \cdot B \cdot C = (A \cdot B) \cdot C = A \cdot (B \cdot C)$	$(A + B) + C = A + (B + C) = \cdots$
Distributive	2(a)	$A \cdot B = B \cdot A$	$A + B = B + A$
	3(a)	$A \cdot A = A$	$A + A = A$
	4(a)	$A \cdot 1 = A$	4(b) $A + 1 = 1$
	5(a)	$A \cdot 0 = 0$	5(b) $A + 0 = A$
	6(a)	$A \cdot \overline{A} = 0$	$A + \overline{A} = 1$
Repeated terms	7(a)	$A \cdot (B + C) = A \cdot B + A \cdot C$	7(b) $A + B \cdot C = (A + B) \cdot (A + C)$
	8(a)	$A \cdot (\overline{A} + B) = AB$	$A + \overline{A} \cdot B = A + B$
De Morgan	9(a)	$\overline{A \cdot B \ldots} = \overline{A} + \overline{B} + \ldots$	9(b) $\overline{A + B + \ldots} = \overline{A} \cdot \overline{B} \ldots$

Fig. 7.7. Logic identities.

ON or OFF, OPEN or CLOSED, HOT or COLD, FAST or SLOW. For some of these examples there is clearly no third state, but it is assumed for *all* the signals being dealt with in combinational logic that there are only two states. With one family of gates, any voltage level from about 2 V up to the 5 V supply to the circuits is taken as being HIGH or 1, while any voltage between zero and 0.8 V is taken as being LOW or 0. From this assumption comes one of the great benefits of digital circuits. Noise or spurious signals picked up on the input wiring will not cause faulty operation provided this noise is below the levels at which the circuit switches from one of its states to the other.

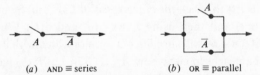

(*a*) AND ≡ series   (*b*) OR ≡ parallel

Fig. 7.8. Switch arrangement to simplify $A \cdot \overline{A}$ and $A + \overline{A}$.

The identities or laws of Boolean algebra are in groups – the first two lines show the *associative* and *distributive* laws. The next group of four lines in fig. 7.7 shows that a variable simplifies if it is in some function interacting with itself, or with 1 (HIGH) or with 0 (LOW), or with its complement. Some people find these laws as obvious as the first two; but those who do not may find that making a quick sketch of a switch arrangement that is identical to the Boolean statement may help. Fig. 7.8 shows how Law 6, $A \cdot \overline{A} = 0$ and $A + \overline{A} = 1$ can be checked. Fig. 7.8(a) shows how the AND arrangement is made; here the switches are in series so only when both the first *and* the second switch are closed is the output joined to the input. As the switches corresponding to A and $\overline{A}$ are always defined as being in *opposite* states, one or the other switch is open giving no output

221

A	B	$A \cdot B$	$\overline{A \cdot B}$	$\overline{A}$	$\overline{B}$	$(\overline{A} + \overline{B})$
0	0	0	1	1	1	1
0	1	0	1	1	0	1
1	0	0	1	0	1	1
1	1	1	0	0	0	0
			LHS			RHS

Fig. 7.9. Truth table illustrating De Morgan's law, $\overline{A \cdot B} = \overline{A} + \overline{B}$.

so $A \cdot \overline{A} = 0$. The reader is left to check the equivalent OR relation shown in fig. 7.8(*b*). (OR is sketched using switches *in parallel*.) It is more important to remember that Boolean statements which look like those in this group can be simplified, and to be able to do this quickly, than it is to memorise all the laws.

Laws 7 and 8 relate to repeated terms in a Boolean statement. Law 7(*a*) may be obvious using the 'multiplying out' analogy in ordinary algebra but 7(*b*) is far from obvious. A switch arrangement is one way of checking its truth. Alternatively the perceptive reader may have noted that all the laws in the left-hand column are the *duals* of those in the right column. That is, an AND replaces an OR, and OR replaces an AND, a 0 replaces a 1 and a 1 replaces a 0. (Laws 4 and 5 may not appear to be duals but 4(*a*) is the dual of 5(*b*) and 5(*a*) of 4(*b*). Note that 7(*b*) can be remembered as the dual of 7(*a*) since the dual of A AND (B OR C) is A OR (B AND C).)

The reader is left to check the correctness of Law 8(*a*) by using Law 7(*a*) (multiplying out) and then using Laws 6(*a*) and 5(*b*). While it appears possible that $A \cdot (\overline{A} + B)$ will simplify – a term repeated with its complement – it may not be nearly so clear how $A + \overline{A} \cdot B$ in Law 8(*b*) would simplify. Remember the idea of duals, or try a switch arrangement or even a Venn diagram to check any doubts.

Lastly, line 9 of fig. 7.7 is called De Morgan's Law. In practice, it gives a way of changing any AND term in a Boolean expression to OR (as in 9(*a*)) or vice versa. Law 9(*a*) can be written: 'the complement of two variables which are ANDed is equal to the complement of the first variable OR the complement of the second variable.' Law 9(*b*) is its dual. Many people may find it helpful to remember De Morgan's law in this verbal form.

It is useful to check Law 9(*a*) using a truth table as an alternative to a switch arrangement. This is shown in fig. 7.9. The possible states that the variables A and B take are listed on the left and the two terms $\overline{A \cdot B}$ and $\overline{A} + \overline{B}$ are built up and are seen to be identical (with practice one would write many fewer lines in the table). The reader who has not met this before is recommended to look back at figs. 7.3(*b*) and fig. 7.4(*b*) where the basic AND and OR tables were introduced.

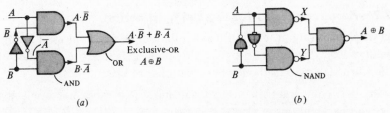

(a)   (b)

Fig. 7.10. Circuit to be simplified: (a) containing a mixture of gates,
(b) with NAND gates only.

Lastly let us return to the reason why we use De Morgan's law. Consider the arrangement of gates shown in fig. 7.10(a). The signals on the left are inputs and by inspection, going from left to right, those on the other lines can be seen to be as marked. So the whole arrangement performs the Exclusive-OR operation and gives an output when the inputs are different!

In practice only *one* type of gate may be available in a single chip package which might contain four gates of the 2-input type. One cannot buy a single package of assorted inverters and OR and AND gates; and to arrange a circuit compactly that uses just a few gates out of several packages makes the layout complex if the remaining gates are to be used in other circuits.

Now there are two solutions to the problem of using the available package to construct the circuit. One is to use AND gates with what is called an 'open collector output'. This is like having a simple transistor switch with no load resistor between the collector and the + supply. By joining together the collector (output) of each gate together and adding a single resistor to the + supply, the OR function is achieved at very little cost. The second solution is to use De Morgan's law; thus Law 9(b) removes the OR function and gives

$$\overline{(A \cdot \bar{B}) + (B \cdot \bar{A})} = \overline{(A \cdot \bar{B})} \cdot \overline{(B \cdot \bar{A})}$$

where $(A \cdot \bar{B})$ is treated as one term and $(B \cdot \bar{A})$ as another. Taking the complement of both sides gives

$$A \cdot \bar{B} + B \cdot \bar{A} = \overline{\overline{(A \cdot \bar{B})} \cdot \overline{(B \cdot \bar{A})}} = \overline{X \cdot Y}.$$

The left-hand side is the required Exclusive-OR expression and the right-hand side, which is equivalent, might at first appear complex until it is realised that it is the NAND function of two terms X and Y and each of these is the NAND of the inputs. So an equivalent circuit is shown in fig. 7.10(b).

223

7.7 Karnaugh maps for logic simplification

In the engineering problem of making the simplest and cheapest arrangement of logic circuits to do a given job, Boolean laws are usually aided by the Karnaugh map method, whose features are that:

(*a*) it is comparatively easy to remember (for those not doing logic design too often, this is a great benefit),

(*b*) designs are simplified by eye,

(*c*) circuits which are free from generating spurious short pulses can be made (called hazard-free design and explained in § 7.8),

(*d*) easy analysis is possible for circuits with up to 4-inputs; the analysis of 5- or 6-input circuits is also possible but more difficult.

The Karnaugh map will be explained using a particular example; namely:

A system to prevent a motor from overheating uses three fans $A1$, $A2$, $A3$. If $A1$ and $A2$, or $A3$ fail, when the motor is running, the circuit is to give an alarm signal L. Draw a map for the logic required and show that it can be simplified if a 'don't care' state exists for fan $A2$ failing when $A1$ is working as $A1$ will provide enough cooling on its own.

(*a*) *Step 1. Define the inputs.* Say that the logic available from transducers on the fans have value 1 if the associated fan is running and 0 if it is not. Also a transducer is attached to the motor to give an output $B = 1$ if the motor is running. Then the output L, to given an alarm, will be written as

$$L = (\overline{A1} \cdot \overline{A2} + \overline{A3}) \cdot B$$
$$= \overline{A1} \cdot \overline{A2} \cdot B + \overline{A3} \cdot B, \text{ by algebra.}$$

Note that one needs to read the logic statement very carefully.

(*b*) *Step 2. Draw a 4-input map.* The Karnaugh map is drawn with the possible input combinations *around* it as in fig. 7.11. Note that they are *not* in the order of the binary count but in the 00, 01, 11, 10 sequence where only 1 bit changes at a time. The sixteen squares inside the map (rather like the sixteen lines of a truth table for four variables) correspond to each of the different combinations of the inputs; for instance, $A1 \cdot \overline{A2} \cdot A3 \cdot \overline{B}$ which is a 4-*variable combination* is mapped to one square at the bottom right corner of fig. 7.11. This is because the first two terms, $A1 \cdot \overline{A2}$, map to the right-hand column of the figure and the last two terms, $A3 \cdot \overline{B}$, map to the bottom row of the figure: where these two intersect is the square mapped by $A1 \cdot \overline{A2}$ AND $A3 \cdot \overline{B}$ which is what we want. Next consider a 3-variable term that we want to map $\overline{A1} \cdot \overline{A2} \cdot B$. The $\overline{A1} \cdot \overline{A2}$ inputs

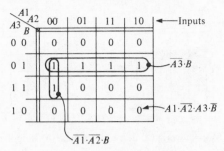

Fig. 7.11. Karnaugh map.

are shown at the top of the left-hand column (both inverted or low). It doesn't matter what $A3$, the first variable on the left is, but $B = 1$ maps the expression into the middle two squares as shown in fig. 7.11. So a *3-variable combination* maps to two *squares*.

By similar reasoning $\overline{A3} \cdot B$ maps into the area shown and it is seen that a *2-variable combination* maps to *four squares* (i.e., it does not matter what $A1$ or $A2$ are). Lastly a *1-variable term* would map to *eight squares*; e.g. $A1 = 1$ would map just into the right-hand half of the map.

By convention the squares mapped are marked 1 or shaded and those unmapped are marked 0 or left clear.

(*c*) *Step 3. Simplify by identifying 'blocks' mapped.* If several of the terms mapped add up to form a large block, say four or eight squares together in a rectangle, then applying the converse of the method of mapping they simplify into a 2- or 1-variable term respectively. So the *larger* the block identified, the *simpler* the gate that is needed to combine the fewer variables to give the logic.

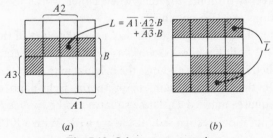

(*a*)   (*b*)

Fig. 7.12. Other map conventions.

Fig. 7.12(*a*) shows the area mapped for the required function L; note the alternative way of showing the variables round the edge of the map which is liked by many designers and the authors. Note that $A3$ is shown to the left of the map so that any term containing $A3$ maps to the bottom

225

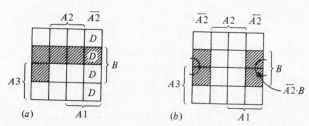

Fig. 7.13. Map showing (*a*) don't care states, (*b*) simplification of $\overline{A2} \cdot B$.

half of the map and any term containing $\overline{A3}$ maps to the top half. By also noting that B maps to the middle two rows (it is shown on the other side of the map), so that $\overline{B}$ is at the top and bottom of the map, the input definition is the same as in fig. 7.11 but it is much quicker to write. $A1$ and $A2$ are shown at the top and bottom of the map.

In fig. 7.12(*b*), $\overline{L}$ has been mapped. If it had been less than half the area, or some simple shape, or containing some term that is already required for another logic purpose, then it might be advantageous to construct a circuit to give $\overline{L}$ and follow it by an inverter. To map to areas that 'add up', then the terms are 'orred' together: the reader may care to check that $\overline{L} = \overline{B} + A3 \cdot A2 + A1 \cdot A3$. It does not matter that some squares are mapped more than once.

(*d*) *Step 4. Further considerations.* Two squares which are touching but on a diagonal are mapped by an Exclusive-OR gate. The reader is left to check this. Finally, the constraints of how many gates come in each package also feature in a designer's choice – thus a design needing a single inverter and having a spare 2-input NAND or NOR gate in a package can be made by joining the gates' two inputs together when either works as an inverter (other connections are also possible).

Now let us finish the example which was set at the start of this section: it is asked 'if $A2$ failing while $A1$ works is a "don't care" state, can the logic be simplified?' Fig. 7.13(*a*) shows the original map and $\overline{A2} \cdot A1$, which is equivalent to the new condition, maps to the right-hand column as shown by the squares marked D. These squares can be *included* or can be *left out* of the conditions recognised, just as we wish. With a little practice, a designer will recognise a new 4-square block being formed because the two outside columns are recognised by $\overline{A2}$ and the centre two rows are recognised by B. So $\overline{A2} \cdot B$ is shown separately in fig. 7.13(*b*) in the shaded squares. It can be considered that the map could be rolled round so that the edges touch – all adjacent squares in the map just differ by a single variable and $X \cdot Y + X \cdot \overline{Y} = X$ is the equation that supports the statement that two neighbouring squares give a shorter logic expression than either on its own. (The reader is left to consider how $\overline{B} \cdot A1$ and $\overline{B} \cdot \overline{A2}$ appear – the latter maps to the four corner squares which again become a 2-variable logic expression!) So the new expression wanted for L is $\overline{A2} \cdot B + \overline{A3} \cdot B$. Algebra and De Morgan can be used to simplify it if it is to be built with the fewest NAND gates as

$$L = (\overline{A2} + \overline{A3}) \cdot B = (\overline{A2 \cdot A3}) \cdot B$$

which uses only NAND or AND functions.

An example is given at the end of the chapter which completes a design by putting it into gate form.

†7.8 Hazards and their elimination (using the Karnaugh map)

A hazard is the production of a *spurious* short pulse at the output of a logic circuit just as the inputs are changing. If the logic is followed by a faster counter, then a wrong count will result as the short pulse will register as an added event.

Consider the logic shown in fig. 7.14(a) (ignoring the broken part of the circuit) which yields an output $X = \overline{A} \cdot B + A \cdot C$ mapping to the squares shown shaded in fig. 7.14(b). Note the simpler map which is needed for three variables. The terms map to areas touching but *not* intersecting with each other; this is the condition which gives a hazard. To explain it, consider the voltage levels A, B, C plotted against time in fig. 7.14(c). A and C are both HIGH at the start so that the lower AND gate will give an output which will be passed through the OR gate to give X HIGH. At a time t_1, A changes from HIGH to LOW. It will take a finite time, perhaps only a fraction of a microsecond, for the inverter to change its output which is shown as the period between t_1 and t_2 on the figure. For this time, both the AND gates will be getting one LOW input and so neither will give an output, and the output as shown will fall briefly to LOW. After t_2, the correct output is again generated.

How is this hazard overcome? First it is noted that it only occurs when both B and C are HIGH on the changeover of A. Thus $B \cdot C$ must be recognised by another simple 2-input AND gate and be corrected as shown 'broken' in fig. 6.14(a). The term $B \cdot C$ now *intersects the two areas* which were mapped before – this is the criterion to be used to make the logic hazard free.

A designer does not always have to add such gates. It is only necessary if the circuit output goes to *fast* acting memory or fast counting circuits.

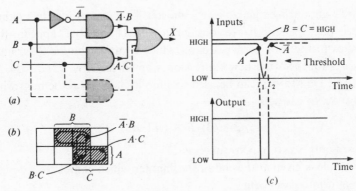

Fig. 7.14. Hazards: (*a*) circuit, (*b*) map, (*c*) voltage–time waveforms.

Most outputs such as alarms, warning lights, actuators, heaters would not have the speed to respond to such a short event. So the expense of the added gate may often be avoided.

7.9 Memory – two inverting gates with feedback

If two inverting gates are connected in sequence with a signal from the output of the second gate being taken back as an input to the first, as shown in fig. 7.15(*a*), then the simple relations of Boolean algebra will indicate conflicting conditions for the output. The circuit becomes a memory. This is a fundamental step which develops a logic circuit into one which can compute by remembering past events. Intercomparison of past and present events is the basis of all decision making in computing. Memory is also a necessary start to counting.

In the circuit of fig. 7.15(*b*) the two NAND gates again have feedback and this is the way that memory or bistable circuits are usually constructed. A bistable is a circuit which has two separate stable states which can be maintained indefinitely in time. A monostable has only one such state and an input makes it change to another state; after an interval, it relaxes to its stable state again. An astable circuit also has two possible states but it is continually interchanging states between one device which is on and another which is off: it is often called a flip-flop. In fig. 7.15(*b*), the two inputs are *S* and *R* and the outputs are Q_1 and Q_2. All other lines are marked with the signals expected. Using De Morgan's law the outputs are:

$$Q_1 = S + \bar{Q}_2,$$
$$Q_2 = R + \bar{Q}_1.$$

228

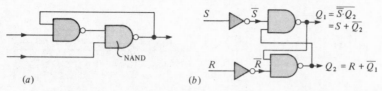

Fig. 7.15. Memory circuits: (*a*) basic two inverting gates with feedback, (*b*) Set–reset bistable.

The equations show that the outputs are *opposite* to each other if $S = R = 0$. If $S = 1$ then Q_1 becomes 1, even if it hasn't been before, from the first equation. The second equation then defines Q_2 as 0 and now S can revert back to zero as $Q = 1$ in the upper equation is now maintained by the second term, namely $\bar{Q}_2$, which is the memory action. (This can be considered in other ways also.) S is called the Set input as S HIGH sets Q_1 HIGH. A similar analysis shows R HIGH makes Q_2 HIGH followed by Q_1 LOW; so R is called the Reset input.

The $S = R = 1$ state gives a complication since it defines both outputs to be high: when $S = R = 0$ again it is *not* defined whether Q_1 or Q_2 stays high. It depends on the speed of the two NAND gates (which is pure chance) and the faster gate will revert to 0 while the slower one will stay high. This is an input condition to be avoided (as illustrated in the circuits described in § 8.2).

†7.10 Realisation of gates with MOS or bipolar devices – 3-state or open collector devices

CMOS (complementary MOS) and TTL (transistor–transistor logic, made with bipolar devices) are the most popular logic families. Fig. 7.16(*a*) shows a CMOS AND gate and a brief consideration of its operation will help the reader to understand its main performance features which are described in the next section. All the transistors are enhancement mode devices and fig. 7.16(*b*) shows a cross-section of the silicon integration. The gate has to overlap the source and drain connections to give a low resistance channel when the gate voltage exceeds about half the supply voltage: this unfortunately gives appreciable input capacitance to the circuit which limits its speed.

When both the inputs A and B are HIGH (or greater than half the supply voltage), T_1 and T_2 will both be able to conduct whereas the complementary devices T_3 and T_4 will not conduct. Any charge existing at X is promptly discharged so X is LOW; this makes T_6 conduct and T_5 not

229

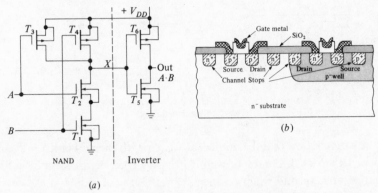

Fig. 7.16. (*a*) CMOS AND gate, (*b*) cross-section of metal gate CMOS inverter.

conduct so a HIGH output is given. Next consider either of the inputs A or B to be LOW so that either T_3 or T_4 will conduct. As they are in parallel, X will be charged to the supply potential (T_1 and T_2 are in series and one of them is high resistance). X HIGH will define a LOW output. So the circuit is an AND gate.

The devices in the circuit act as switches. The connection from the supply to ground can be traced through an ON transistor in series with an OFF transistor in all the modes mentioned. So *no steady current* is drawn by the circuit in either of its usual states but there is a pulse of current during each changeover. It charges up the point X or any output loads. So a CMOS circuit runs with no supply power taken when static but the rate of charge taken increases with the rate of switching up to a level close to that of TTL circuits at the maximum switching rate of the circuit. This is why calculators with CMOS circuits are switched to a standby mode to save power if no entry is made, say for 30 seconds.

The circuit of a simple TTL NAND gate is shown in fig. 7.17. The input npn transistor T_1 has a double emitter which achieves the AND function, T_2 inverts the signal and T_3, T_4 form the output stage in which one or other of the two transistors conducts. When T_3 conducts, the output becomes about 3.6 V (5 V less the voltage drop in two diodes). However, if no input circuit is connected, no current flows in the inputs but a path from 5 V through the 4 k base resistor takes the base–collector p–n junction into conduction which is followed by the base–emitter junctions of *both* T_2 and T_4. So T_4 is turned on; T_2 which is passing current through the 1.6 k load makes the base voltage of T_3 low and the added diode in series with the emitter of T_3 keeps it off. So floating inputs are equivalent to both inputs being HIGH; they define a LOW output.

230

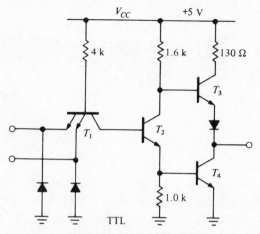

Fig. 7.17. TTL NAND gate.

When both inputs are HIGH (at say 3.6 V or more), T_1 with its n 'emitter' regions more positive than its n-type 'collector' region acts in an inverted mode (of low gain only as the lightly doped collector is not a good emitter). Because of the appreciable base current through the 4 k resistor, current is passed into T_2 and T_4 bases as before so again the output is LOW.

Lastly when either one (or both) of the inputs is LOW, T_1 will act in a normal mode – the emitter at ground and base current through 4 k makes the device conduct as needed to discharge the base of T_2 to a low potential of 0.7 V or less. This is not enough to turn T_2 on: the 1.6 k collector resistor of T_2 now only feeds base current to T_3 to turn it ON and this makes the output HIGH. This can be seen to be the NAND operation.

Logic circuits can be bought with a different output circuit called open-collector or 3-state output. In § 7.6, it was mentioned that the OR function could be achieved by having such gates which have essentially TTL circuits as in fig. 7.17, but with T_3 omitted; or CMOS circuits, as in fig. 7.16, but with T_6 omitted. Several gates can then be connected together and an *added* resistor of the order of 1 k to the + supply will complete the 'wired-OR' circuit.

The use of such circuits is fundamentally necessary in 'bus driving'. (A bus is a group of wires which can be 'written to' or 'read from' by many circuits connected to it and with only one device being 'enabled' to 'write to' a line at any time and so to define its voltage.) The inexpensive, practical way to do this is with a wired-OR circuit but it has the disadvantage of slow speed as the finite resistor only charges up the line exponentially to high after being low. This is slower than a conducting transistor. A TTL package

231

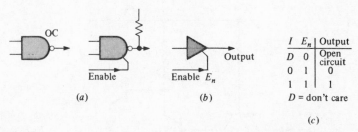

I	E_n	Output
D	0	Open circuit
0	1	0
1	1	1

D = don't care

(c)

Fig. 7.18. Open collector 3-state gates.

to do this would have the circuit of fig. 7.17, with T_3 omitted, as mentioned, and with an extra emitter connection on T_1 labelled 'enable'. With this input held LOW, the other inputs are ignored and T_2 and T_4 are off; then other gates could define the output level. With the enable input HIGH, the gate would work normally. Open collector gates may have the usual gate symbol with an asterisk or OC beside them as shown in fig. 7.18(a).

The same effect can be achieved better by 3-state logic. The name is misleading and what it means is that the output is HIGH, LOW, or floating so it can be defined by other circuits connected to it. Such circuits have both transistors in their output circuit so, when 'enabled', the output can be defined HIGH or LOW at full speed. Fig. 7.18(b) and (c) show their symbols and a truth table. The reader interested in the circuit details should look at more advanced texts.

7.11 Comparison of logic families

The advanced texts and journals have tables and graphs comparing the features of all types of logic. With new types of circuit design and device construction being introduced every year or two, these go quickly out of date.

TTL (transistor–transistor logic) or CMOS (complementary MOS) should satisfy most users' needs except where the ultimate in speed is needed when emitter-coupled logic (ECL) should be chosen. Many types of TTL circuits are available and at present the best type for most applications is the low-power Schottky variety specified by the letters LS appearing somewhere in the code for the circuit, e.g. 74LS00 is a package containing four 2-input NAND gates, and many makers use this code. CMOS gates are now available in a 74C...series which are functionally similar to the TTL series and these, and the 4000 series from RCA, the original makers of CMOS, are the main circuit ranges.

232

The features of TTL and CMOS may be summarised as follows: TTL is the best choice where speed is important. CMOS imposes fewer restraints on the power supply (3 V to 18 V at very low currents) and it has a good noise rejection when a supply voltage of 10 to 18 V is chosen, the threshold between HIGH and LOW can be at half the supply level.

In more detail, the differences between the two families of gates are:

Supply voltage. TTL requires $+5$ V ± 5 per cent. CMOS operates with a wide voltage range, of which $+5$ V and $+12$ V are popular choices and the current taken is much lower.

Inputs. A TTL input must be *held* LOW when typically 1 mA flows *out* of the gate circuit into whatever input is connected to ground. This is satisfactory when one TTL gate is wired to another since each output is good at sinking current. The TTL logic threshold is equivalent to the voltage across two diodes (about 1.4 V) above ground. CMOS, in comparison, draws no input current and has a threshold at about half the supply level. It is normally bought with its pins pressed into conducting foil or conducting foam as the circuit can be damaged by static charge picked up during handling or from a defective soldering iron. Both TTL and CMOS should have unused inputs connected either to high or to low to avoid spurious signal pickup and not left floating.

Outputs. TTL output is a saturated transistor connected between supply and ground in the LOW state where it can sink several tens of milliamps. It is an emitter follower in the HIGH state with a level about two diode voltage drops below the 5 V supply. There are problems in interfacing TTL to CMOS circuits because of the low value of the HIGH level. CMOS output is an ON MOSFET either connected to ground or to the supply in its LOW and HIGH states respectively. This symmetry of operation helps in noise rejection. Both TTL and CMOS outputs can drive at least ten inputs of similar circuitry. This is termed a 'fan-out' of ten.

Speed and power. TTL is faster. The LS series of gates have a maximum propagation delay of 15 ns, and have an average current consumption per gate of 0.4 mA. The figures are 5 ns and 4 mA/gate respectively for the faster S (Schottky) series. CMOS gates have a typical propagation time of about 40 ns but have negligible quiescent power dissipation. However, the power taken increases with increasing frequency of input and CMOS dissipates nearly as much as TTL at its upper frequency limit.

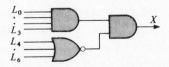

Fig. 7.19. Circuit for worked example 1.

7.12 Worked examples

7.12.1 Analysis of address decoder

A microprocessor uses eight address lines, L_0 to L_7 in a bus to define its memory locations. Check that the logic circuit, shown in fig. 7.19, connected to the lines will give an output X to detect the two addresses 1111 0000 and 1111 0001 (L_0 first and L_7 last).

Solution. By inspection of fig. 7.19, we can write down $(L_0 \cdot L_1 \cdot L_2 \cdot L_3)$ as the signal out of the upper AND gate and $(\overline{L_4 + L_5 + L_6})$ as the signal out of the NOR gate. Hence we can write

$$\text{Output, } X = (L_0 \cdot L_1 \cdot L_2 \cdot L_3) \cdot (\overline{L_4 + L_5 + L_6})$$
$$= L_0 \cdot L_1 \cdot L_2 \cdot L_3 \cdot \overline{L_4} \cdot \overline{L_5} \cdot \overline{L_6} \text{ (using De Morgan's Law)}$$

Note that in the addresses we needed to detect, the condition of L_7 did not matter so the logic is suitable.

7.12.2 Synthesis of a month-length detector

A binary counter having four output lines D_3, D_2, D_1 and D_0 (D_0 least significant) is set up to count the months of the year with D_3 to D_0 = 0001 for January, 0010 for February and so on. What logic using NAND gates and inverters could be connected to the four counter lines to give an output S for short months of the year of less than 31 days? The counter will be reset after 1100 for December to 0001 for January so higher counts and 0000 are 'don't-care' states (and can be either included or excluded on a map in making simple blocks).

Solution. First we must list the input codes that must be recognised on the four counter lines; these are shown in fig. 7.20(*a*). One by one the short months are put into the map fig. 7.20(*b*). Thus 0010 for February is in the top right corner and is marked 2 (the decimal for short), April is marked 4, etc. Then 0000 and the other 'don't care' squares are marked D. On the map, squares 9 and 11 with the two D squares above them yield a simple block. Noting the labelling round the map which is for when each state is high, this square is $D_3 \cdot D_0$. Next the 2, 4, 6 squares and the top left D square yield another simple block, and $\overline{D_3} \cdot \overline{D_0}$ maps into the top half of the map ($\overline{D_3}$) and into the outside columns ($\overline{D_0}$), and is the required term.

234

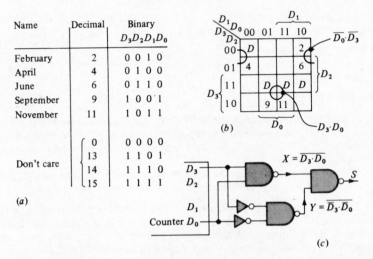

Name	Decimal	Binary $D_3D_2D_1D_0$
February	2	0 0 1 0
April	4	0 1 0 0
June	6	0 1 1 0
September	9	1 0 0 1
November	11	1 0 1 1
Don't care	0	0 0 0 0
	13	1 1 0 1
	14	1 1 1 0
	15	1 1 1 1

(a)

Fig. 7.20. Solution of worked example 2: (a) list of inputs, (b) map, (c) circuit.

Lastly, De Morgan's law gives the expression for S, the short months, as:

$$S = D_3 \cdot D_0 + \overline{D}_3 \cdot \overline{D}_0 = \overline{(\overline{D_3 \cdot D_0}) \cdot (\overline{\overline{D}_3 \cdot \overline{D}_0})} = \overline{X \cdot Y} \text{ etc.}$$

This is made using NAND gates repeatedly as shown in fig. 7.20(c). Note that only two lines from the counter are used. It may be of interest to note that a single gate, known as a parity gate, would do exactly what is required to yield S – it has two inputs and gives a HIGH output when both its inputs are HIGH and when both its inputs are LOW.

7.13 Summary

A digital signal can either be an output on a single wire from some logic arrangement that detects a required set of conditions, or a group of wires (a bus) can carry a set of signals, perhaps arranged in a binary code, to represent an analogue signal. Expressions are written in Boolean algebra to express logic statements, and the statements are simplified with the algebra or with a map by designers of systems. An alternative approach is to use particular large scale integrated circuits (LSI) (see next chapter) to give economic solutions.

Two inverting gates with overall feedback give the very special property of memory in electronics. The circuit is called a bistable circuit and its various applications follow in the next chapter.

A designer of circuits has the choice of using CMOS or TTL or other

235

families of gates. The main features of these are compared in the chapter and capabilities of future types of circuits should be added by the reader to continue the comparison.

7.14 Problems

1. (*a*) A logic circuit has four inputs A, B, C, D and one output X. The output is to be at logical 1 if any three or all four inputs are at logical 1. For other conditions the output is to be 0. Devise a logic circuit using simple AND and OR gates which satisfies this requirement.

(*b*) What is the simplest way to satisfy the requirement if either NAND gates only or NOR gates only are to be used?

2. (*a*) The circuit in fig. 7.21 (*a*) does not make efficient use of logic gates. Write a Boolean expression for Z and hence show how Z can be realised more simply.

(*b*) Fig. 7.21(*b*) shows an arrangement of switches and a relay with two contacts D. Write conditions for the output to be equal to the battery volts; hence qualitatively explain how this circuit contains an element of memory.

Write a Boolean expression for the output and, by using De Morgan's law, show how the logic can be made with NAND gates only. (In this arrangement you should have feedback across two stages of inverting gates which gives memory.)

3. (*a*) Simplify the Boolean expression

$$Z = (A + \bar{B} + C) \cdot (A + \bar{B} + \bar{C}) \cdot (A + B + C)$$

How can Z be realised from inputs A, B, C (but not their complements) using a single quadruple 2-input NAND gate?

(*b*) Using a Karnaugh map, write simple sum-of-products expressions for F and $\bar{F}$, where

$$F = \bar{A} \cdot \bar{D} + \bar{B} \cdot \bar{C} + \bar{A} \cdot B \cdot \bar{C} \cdot D$$

and $A \cdot \bar{B} \cdot C \cdot \bar{D}$ is a 'don't-care' state.

4. (*a*) Using a Karnaugh map, write simple sum-of-products expressions for F and $\bar{F}$, where

$$F = \bar{A} \cdot \bar{D} \cdot B + B \cdot \bar{C}$$

and $A \cdot B \cdot C \cdot \bar{D}$ is a 'don't-care' state.

(*b*) Fig. 7.22 shows a circuit (the half-adder) for adding two binary digits A and B. Express the sum S and carry C as functions of A and B. Using De Morgan's law or otherwise show how this circuit can also be implemented in NAND gates only. Draw a diagram of such an implementation.

Show how two circuits of this type can be combined with a 2-input OR gate to form a full-adder circuit with inputs A, B and C (carry-in) and outputs S and C_0 (carry-out).

(Cambridge University: Second year)

5. For the circuit shown in fig. 7.23, write a Boolean expression for the function F and minimise it using a Karnaugh map. What arrangement could be used to realise F using NAND gates only?

What are the general advantages to be gained in making a logic circuit with one type of gate only?

(Cambridge University: Third year)

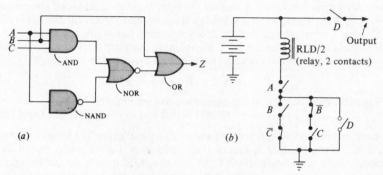

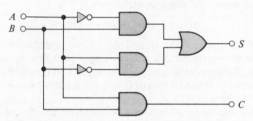

Fig. 7.21. Circuits for problem 2.

(a) (b)

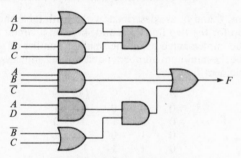

Fig. 7.22. Circuit for problem 4.

Fig. 7.23. Circuit for problem 5.

6. (a) Construct truth tables for the following expressions

(1) $X = A + \bar{B}$
(2) $Y = \bar{A} \cdot B$

Hence state the relationship between X and Y.

Using the basic laws of Boolean algebra show that

$$B \cdot (A + \bar{A}) = B + A \cdot \bar{A}.$$

By plotting the left-hand side of this expression on a Karnaugh map, state and demonstrate the principle of Karnaugh map reduction.

237

(*b*) An output function Z is true for the binary-coded-decimal (BCD) numbers 2, 4, 5, 6, 8, 9, 10, 11, 13 or 15. Design the minimum circuit for the realisation of Z using 3-input NOR gates only, (complemented inputs are available).

(*c*) Explain what is meant by a hazard in a combinational logic circuit and determine the conditions under which a hazard exists in the expression:

$$Y = \overline{A \cdot B \cdot C} + \overline{A} \cdot C \cdot D + A \cdot B \cdot \overline{D}$$

Suggest a modification to the expression to eliminate the hazard.

(Royal Naval Engineering College: Third year)

7. Each of four people, members of a television panel game, has an off–on switch (A, B, C, D) that is used to record their opinion of a certain pop record. It is required that a scoreboard shows a 'hit' when a majority is in favour by lighting a lamp H, and a 'miss' when against by lighting a lamp M. Provision must also be made to indicate a 'tie' by lighting a lamp T.

Explain and prove the Boolean statement

$$H = A \cdot B \cdot (C + D) + C \cdot D \cdot (A + B).$$

Obtain a similar expression for M, and show how these statements can be changed to use the NAND operation only. How many packages would be required to realise H and M if only NAND gates are available and if each package can contain either two 4-input or three 3-input or four 2-input gates? Assume that the switches are *not* wired to give signals $\overline{A}$, $\overline{B}$, etc.

Explain how the tie condition can be obtained easily and show what NAND gates are needed to light the lamp T.

(Cambridge University: Third year, part question).

8. Two functions, f_1 and f_2, are described by the truth table shown below. Draw a Karnaugh map for the two functions and show on a circuit diagram how the functions may be implemented using the smallest number of NAND gates only.

What would be the minimum number of quadruple 2-input NAND gate packages required to implement f_1 alone?

A	B	C	f_1	f_2
0	0	0	0	1
0	0	1	0	0
0	1	0	1	1
0	1	1	1	1
1	0	0	1	1
1	0	1	1	1
1	1	0	0	0
1	1	1	0	1

(Cambridge University: Third year, part question)

8

Digital circuits and applications

8.1. Introduction

A simple memory element, made of two inverting gates and also called a bistable circuit, leads to applications such as switch debounce circuits and counters are made of memory elements suitably connected. A counter, apart from its clear application in counting events, data logging and frequency metering, can lead to a powerful way of converting analogue signals into a digital code. This is analogue to digital conversion (ADC) and is a useful 'interface' in applying digital circuits to the wide field of measurements.

The memory, or bistable circuit, suitably connected, also forms accumulators and registers on which a computer does its arithmetic. In this chapter, the principles are explained but, for details, the reader is referred to texts on microcomputing or computer science. One important application of a register is to convert a many-bit code into a voltage waveform suitable for passing down a single pair of wires as a cheap form of communication: this is called parallel to serial conversion. A register can also be used to form random signals which can be useful test signals or can simulate noise.

All these applications build on the basic knowledge of gates, Boolean algebra, and the basic memory element.

8.2 Bistable circuit (*S–R* type, latch, *J–K*, *M–S* and clocked types)

The bistable circuit described in § 7.9 was called a Set–Reset (*S–R*) bistable. To understand its mode of operation and to see it in an elegant and useful application as a 'latch', consider the circuit shown in fig. 8.1(*a*). Starting at the input where the switch is shown at the 'set' position, the voltage at point A is 0 and it only goes HIGH when the switch is taken away from the 'Set' position: so the A line connected to the gate is marked $\bar{S}$ to denote

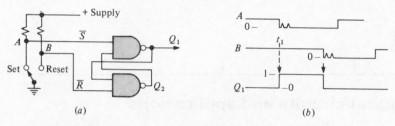

(a)   (b)

Fig. 8.1. Circuit and waveforms of bistable used for switch 'debouncing'.

this inversion that is taking place. Similarly the B line is HIGH when Reset is *not* selected so it is marked $\overline{R}$. Now algebra gives:

$$Q_1 = \overline{\overline{S} \cdot Q_2} = S + \overline{Q_2},$$
$$Q_2 = \overline{\overline{R} \cdot Q_1} = R + \overline{Q_1}.$$

These are identical equations to those derived in § 7.9. In particular, if Q_1 is LOW when the switch is put to Set, the equation for Q_1 makes it HIGH. The equation for Q_2 defines it as the inverse of Q_1, and the second term in the equation for Q_1, namely $\overline{Q_2}$, defines Q_1 to be HIGH even if S disappears. This is exactly what happens in all but special switches. For perhaps a millisecond, the contacts bounce after they first touch so the voltage on line A, Fig. 8.1(b), is first defined LOW as the switch moves to set but will revert to HIGH several times before settling. The output Q_1 takes no notice of those changes and is only reset LOW when the line B is taken LOW and this again is 'debounced'. Note that the switch doesn't bounce back to its alternative position but the contacts become open circuit and the resistors connected to $+5$ V then define the input as HIGH.

This memory capability is used in computer circuits in the 'latch', which is just an S–R bistable as described here. It holds the output, which may be data to be defined on a line, until the next piece of data is to be defined on that line.

A useful variation of the S–R bistable or flip-flop is the clocked type. The circuit is shown in fig. 8.2. Instead of having inverters on the input,

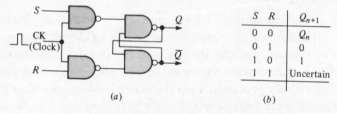

S	R	Q_{n+1}
0	0	Q_n
0	1	0
1	0	1
1	1	Uncertain

(a)   (b)

Fig. 8.2. S–R clocked flip-flop: (a) circuit, (b) truth table.

Bistable circuit

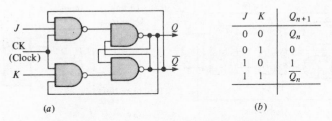

J	K	Q_{n+1}
0	0	Q_n
0	1	0
1	0	1
1	1	$\overline{Q_n}$

(a)　　　　　　　　　(b)

Fig. 8.3. J–K flip-flop: (a) circuit, (b) truth table.

the S and R lines are only connected through to the output gates when the clock input is HIGH. This condition is shown in the truth table where the right-hand column for output is Q_{n+1} meaning *after* the clock pulse before whose start the state was Q_n at the output. Thus $S = R = 0$ is the state for no output change, i.e. $Q_{n+1} = Q_n$: $S = 1$ makes the output set to 1 after the clock and so on. Note that $S = R = 1$ is a combination of inputs which should be avoided. This is done in the following two types of flip-flops which are generally more useful than the S–R type.

The D-type flip-flop is similar to the S–R type of fig. 8.2(a) except that it has only one input D connected to the S or set line. The R line is connected to S through an inverter. When $D = 1$ the output is 1 after the clock pulse: when $D = 0$ the output is 0 after the clock pulse. Hence the name of the circuit is D or data flip-flop as it is used for transferring data from input to output at each clock pulse.

The second improved type is the J–K flip-flop where the condition for both inputs being HIGH is given a useful purpose. Consider the circuit shown in fig. 8.3(a): the input gates now have a connection to the output of the circuit. As before, the inputs, now called J and K, are only 'enabled' when the clock line is high. If, however, both the inputs are HIGH, the cross-connection to the output of the circuit means that the flip-flop will always change state on receipt of a clock pulse – e.g. if Q has been 0, then $\overline{Q} = 1$ will be connected to the upper input gate so, on the reception of a clock pulse, this gate is 'enabled' to set Q to 1. The circuit 'see-saws' or 'toggles'. The truth table in fig. 8.3(b) shows the additional state.

This simple circuit has one major problem – it will only work properly with *very short* clock input pulses to just toggle *once* when $J = K = 1$. A race develops; if the clock input stays high for long, then the output change (after 20 ns or so which is a typical gate delay) is fed back to the input. The opposite input gate now receives all inputs HIGH and so changes its output, and so on. There are three solutions to this problem:

241

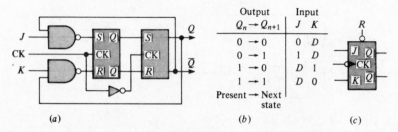

Fig. 8.4. (*a*) Master–slave (*M–S*) *J–K* flip-flop, (*b*) characteristic (excitation) table, (*c*) circuit symbol.

(*a*) to use very short clock pulses, but this is satisfactory only in theory, since wiring capacitances delay these short pulses and timing errors result;

(*b*) to use an edge-triggered flip-flop (see fuller texts for details), but the effect is similar to (*c*);

(*c*) to use a master–slave arrangement: the circuit shown in fig. 8.4(*a*) looks complicated but it is usually available at little extra cost in the common ranges of integrated circuits.

The master–slave circuit works as follows:

(*a*) During the time that the clock input is HIGH, the *J* and *K* inputs set up the first flip-flop. The second flip-flop, whose connection to the clock is through an inverter, does not respond at this time. Note that the input gates again have a 'steering' connection to the output of the whole circuit to resolve the $J = K = 1$ condition.

(*b*) When the clock signal goes LOW again, no more changes can be read into the left-hand flip-flop but the right-hand one now has a HIGH clock and so it becomes a 'slave' and passes on the state of the input flip-flop. This state will be maintained until the next time the clock makes a transition from HIGH to LOW, when again the state of the master (or left-hand) flip-flop will be passed to the output.

Fig. 8.4(*b*) shows a variant of the truth table of fig. 8.3(*b*) which is useful in circuit design. In the left-hand column are listed the four ways that the circuit can respond to a clock pulse, Q_n being the state before and Q_{n+1} being the state after. Looking at the middle two lines first, a 0 to 1 transition is caused by $J = 1$ and it doesn't matter what *K* is; 1 to 0 is caused by $K = 1$: so the circuit works like a Set–Reset circuit with $J =$ Set, etc. The top line of the table shows that for the 0 to remain 0, we need not set the circuit, i.e. $J = 0$ and *K* is again 'don't-care'. Lastly for a 1 to remain 1, we need not reset so $K = 0$. This table will be used often in the designs

242

that follow and it is a most useful variant of the truth table. It is called a transition or excitation table.

Lastly the symbol for the circuit most usually seen in makers' data books is shown in fig. 8.4(*c*). A little circle, meaning inversion, is shown on both the Clock (CK) and Reset (*R*) inputs. Thus the circuit changes its output as the clock goes LOW and the small triangle is a convention for an edge-trigger. The circuit is reset by making $R = 0$; this does not mean that a user can leave it unconnected. It has to be joined to the + supply if the Reset is never used, or connected to 0 V when the output will be reset to LOW regardless of the presence of a clock pulse. With many applications, it is desirable that the circuits all start from the Reset state so a capacitor-resistor circuit is often used to give a short delay. It is connected to all the Reset lines of the flip-flops in an equipment.

8.3 Counters – ripple and synchronous types

Some circuits using flip-flops are simple enough to need little design theory to understand or use them. Such circuits are ordinary counters and registers which are described in this and the next section. However, a design method using 'state diagrams' is important for many designs and this will be described in §8.5.

Most counters use the 'toggling' action of flip-flops. By considering the circuit of fig. 8.5(*a*), where the *Q* output of each flip-flop is joined to the clock input of the next one, each stage makes a 'divide by 2' or binary counter. Note that all the *J–K* flip-flops are used with $J = K = 1$ which causes each flip-flop to change to its opposite state (toggle) as its input clock pulse goes from 1 to 0. The figure shows just the first three stages of possibly many, if a counter for large numbers is needed.

Consider the state of the counter as shown in fig. 8.5(*b*), with

$$Q_0 = Q_1 = Q_2 = 0$$

initially. The input pulse train *I*, which is to be counted, starts; at each change from 1 to 0, the first bistable will be 'toggled' so Q_0 goes 0 1 0 1 as shown. Note that Q_0 is switching at half the rate of *I*; hence the name 'divide by 2' counter is used. But Q_0 is also connected to the next bistable which will toggle (and divide by 2) so Q_1, and later Q_2 are switched on each 1 to 0 transition. Suppose that, at a time t_1, the state of the counter is examined, $Q_2 Q_1 Q_0$ will be 100_2 which is 4_{10}. Note the simplicity of the circuit – it does not have to be wired up from separate *J–K* packages as 4-bit and 5-bit counters are commonly available as integrated circuits with

243

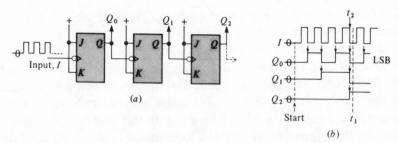

Fig. 8.5. Simple 3-bit counter: (*a*) circuit, (*b*) timing diagram.

added inputs to clear the count to 0 and even to preset in a desired start state. Packages are also available that will count down as well as count up.

The simple counter shown is called a 'ripple' counter. This is because of the way that the bistables respond to input. Consider the time marked t_2 at the end of the fourth pulse on I. A small but finite time will be taken for the change I to give the change Q_0 at the output of the first bistable and then there is a delay until Q_1 changes. In its turn, this change is the right one to make the next bistable toggle and so the change 'ripples' through the circuit. If there are n stages each with propagation time t_p, $n \times t_p$ will be the delay before the correct count is set up. If some logic is waiting to decode the state to signal that the count is complete, then the delay may be serious particularly if the input I is of high frequency (t_p is typically 20 ns): this delay is one of the disadvantages of the ripple counter. The other disadvantage is that spurious outputs can be read if the counter is examined during a 'ripple' of changeover.

The synchronous 4-bit counter, whose schematic diagram is shown in fig. 8.6, overcomes the two disadvantages of the ripple counter. The four output lines and the clock-up input at the top left could be expected on a simple counter. (Note the triangle on the clock input line, but no circle, meaning that the output changes at each edge going HIGH.) The 'carry' output will be signalled when the full counter is resetting back to 0 which allows counter chips to be 'cascaded' to get higher counts. The four 'load' inputs can have any desired binary code connected to them which will be entered regardless of the state of the counter when the load input is taken LOW which allows the count to be started from other than the 0000 state (achieved by Reset). Furthermore, by switching the 'clock input' to the DOWN line, the counter will decrement its state and the 'borrow' output will change when the 0000 state changes to 1111, again allowing stages to be cascaded. A similar chip gives BCD outputs which are useful for decade counts direct to displays.

244

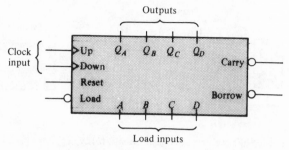

Fig. 8.6. 4-bit synchronous up/down counter (74 LS 193).

A feature often wanted in counters is to make them count to some number other than the usual 10, 16, 64, etc. For example, 12 or 50 may be a desired number to be recognised. There are three ways of doing this, and the methods are described for a binary N-bit counter which would count to 2^n but where an arbitrary number X is to be recognised:

(*a*) Load $(2^n - X)$ into the counter at the start and recognise the counter as full from the 'carry' output: this achieves the desired shorter count but the numbers in the counter do not mean much.

(*b*) Load X into the counter and *count down* and recognise the end by a 'borrow' being signalled: the state of the counter is now quite useful as a display and this is easy to implement.

(*c*) Recognise the count X on the output lines using a suitably designed gate circuit: there is some cost and effort involved in the added design but again the display is useful.

So solution (*b*) has much to recommend it.

8.4 Registers, accumulators and arithmetic ICs

Consider a number of flip-flops, connected as shown in fig. 8.7(*a*) where the outputs of each stage, Q and $\bar{Q}$, are connected to the $J–K$ inputs of the next stage. All the clocks are driven together, so that at each pulse on this line, the state of the register shifts to the right, with data entering from the 'data input' on the left. Thus a waveform, consisting of a series of N HIGH or LOW states, can be 'clocked into' the register and after N pulses it will be set up in the N bistables. This is called serial to parallel transformation, where N bits on *one* line are converted to data on N parallel lines.

Shift registers can also be 'loaded' with N bits of data, one in each bistable and then the data 'clocked-out' at the right-hand bistable; thus

245

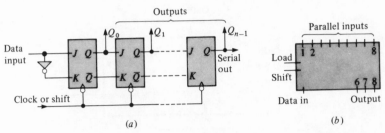

Fig. 8.7. Shift registers: (a) basic circuit, (b) typical 8-bit package.

achieving parallel to serial transformation. The two sorts of shift registers are much used in communication links where data of many bits are to be passed into a single line. Fig. 8.7(b) shows a typical register where an 8-bit code is accepted into the chip when the load line is taken HIGH. It can be clocked out in serial form when the shift line is given eight pulses.

An accumulator is a shift register on which arithmetic can be done. An $(N+1)$-bit shift register with the first bit at 0 and the remaining ones set up with data is one of the conventions for a positive N-bit number. If all the bits are complemented, including the first, and 1 added to the least significant bit (with carry up if necessary), then we have one of the most powerful ways of treating negative numbers. (Computing Science texts should be consulted for more information.) Multiplication by 2 is achieved by shifting a binary number one place up a register towards the more significant end: division is by shifting the other way. An accumulator has a 'load' input line which transfers the data on eight input lines into the register, so wiping out the original state. It also has an 'ADD' input line which takes data on eight input lines and adds it to the data already in the register. Clearly there are 'carry' steps involved, but extra gates in the package organise this so that within a little more time than the time to set up one flip-flop, the correct output is achieved.

A microprocessor contains several accumulators and registers and has the ability to do a wide range of arithmetic on data; all but the simplest processes tend to use a microprocessor. § 8.11 gives an introduction to the uses and capabilities of microprocessors.

† 8.5 State diagrams and sequential design

Some circuits using flip-flops, such as counters and registers, are so simple that no design theory is needed for their understanding and use. 'Sequencers' which generate a number of states in a given order, need some theory to

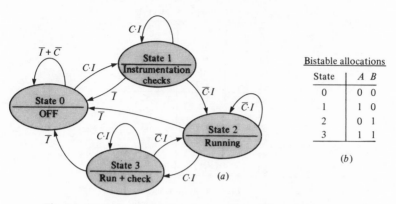

Fig. 8.8. (*a*) State diagram for given problem, (*b*) state allocation.

help in their design – one of the most popular methods is the use of the state diagram.

Consider the design of a circuit to control the sequence for *safely* turning on and off an engine under test. To go from state 0, the OFF state, safety interlocks I have to be complete and a 'check instruments' button C must be pressed: this new state (state 1) has to be maintained as long as the button C is pressed (so that fuel levels, temperatures, oil levels can be read from instruments powered by the circuit). On releasing C, a new state (state 2) will result where the instrumentation is *not* powered but the circuit gives a signal to start and run the motor. If C is pressed again, the motor runs *and* the instruments are energised so that they can be read (state 3); on release of C, the state 2 is resumed. At *any* state, an incomplete interlock circuit, say $\bar{I}$, results in a return to state 0. (An OFF switch can be wired in series with the interlocks.) One can imagine further states for operating a real system, but these few will illustrate the method, which has the following steps:

(*a*) *Draw the state diagram.* This is shown in fig. 8.8(*a*). Each block represents a state, the lines between blocks have a direction marked together with the signals that will initiate the move between states. Thus I AND C are needed to move from state 0 to state 1 whereas $\bar{I}$ will reverse this. State 0 is maintained while C is *not* pressed OR the interlocks completed; state 1 is maintained while $C \cdot I$ exists; and these circular paths out and back to each state are important. The reader can check the other states and the signals on the lines between them for himself. It is important to read the statement of the problem very carefully at this stage so that the right situation is analysed.

247

Present state	Inputs	Next state
S0	$\bar{I} + \bar{C}$	S0
S0	$C \cdot I$	S1
S1	$C \cdot I$	S1
S1	$\bar{I}$	S0
S1	$\bar{C} \cdot I$	S2
S2	$\bar{C} \cdot I$	S2
S2	$\bar{I}$	S0
S2	$C \cdot I$	S3
S3	$C \cdot I$	S3
S3	$\bar{I}$	S0
S3	$\bar{C} \cdot I$	S2

Fig. 8.9. State table.

(b) *Bistable allocation*. If there are X states, then a series of N bistables (like a binary count) can define 2^N states so a number N is required where $2^N > X$. In this problem, two bistables are enough to represent the four states and a possible allocation of the outputs of the bistables A and B is shown in fig. 8.8(b). A little thought has been given to the choice: in state 1, the high output of bistable A can directly energise the instrumentation, and in state 2, A if off and B can directly start the machine by energising a relay with suitable contacts to connect to a starter circuit. Some experience helps to select a set of codes which is economical. For a more complicated problem, it may be better to select a sequence for the bistables which is a binary sequence and to get on with the design. (A little logic may be necessary to decode the states at the end.)

(c) *Drawing a state table*. A state table, sometimes called a transition table, lists each state of the system on the left, its possible next state on the right and, in a centre column, the input conditions associated with the change. Fig. 8.9 shows the table for the example being considered. Again it is important that all the state changes, shown by lines in the state diagram, are entered.

The state table is now expanded for the particular bistable allocation and bistable type to be used. Fig. 8.8(b) gave the bistable allocation so fig. 8.9 can be developed to fig. 8.10(b), where in the left-hand column, SO (state 0) is replaced by 00 for the bistables A and B which are to represent it. In a similar way all the entries in the left-hand column and in the next state column are made (carefully using the agreed allocations).

†(d) *Adding the excitation column to the state table*. Sequencer design is

---

† A first-time student of logic may well skip quickly through this section and the next one looking at the idea rather than the detail.

State diagrams and sequential design

Output $Q_n \to Q_{n+1}$	Input J	K
0 → 0	0	D
0 → 1	1	D
1 → 0	D	1
1 → 1	D	0
Pre-sent → Next state		

(a)

Present state A	B	Input to system	Next state A	B	J_A	K_A	J_B	K_B
0	0	$\bar{I}+\bar{C}$	0	0	0	D	0	D
0	0	$C{\cdot}I$	1	0	1	D	0	D
1	0	$C{\cdot}I$	1	0	D	0	0	D
1	0	$\bar{I}$	0	0	D	1	0	D
1	0	$\bar{C}{\cdot}I$	0	1	D	1	1	D
0	1	$\bar{C}{\cdot}I$	0	1	0	D	D	0
0	1	$\bar{I}$	0	0	0	D	D	1
0	1	$C{\cdot}I$	1	1	1	D	D	0
1	1	$C{\cdot}I$	1	1	D	0	D	0
1	1	$\bar{I}$	0	0	D	1	D	1
1	1	$\bar{C}{\cdot}I$	0	1	D	1	D	0
					J_A	K_A	J_B	K_B

(b)

Fig. 8.10. (a) Excitation table for J–K flip-flop, (b) full state or excitation table for the problem.

most satisfactorily accomplished by using J–K master-slave flip-flops and its excitation table, discussed earlier, is repeated for clarity in fig. 8.10(a). Consider the similarity of the present state, next state and inputs column headings in both figs. 8.10(a) and 8.10(b). D stands for 'don't care' in fig. 8.10(a), so only $J = 0$ is needed for a 0 state to stay 0 after a clock pulse. This is exactly the condition that bistable A has to follow in the top line of fig. 8.10(b) and it is shown shaded together with the input $J_A = 0$, $K_A = D$ which can be written into the right-hand column, again shown in a shaded area. By chance, the present to next state conditions for A in the top four lines of Fig. 8.10(b) are nearly the same as those in the J–K flip-flop excitation table and again they are shown shaded. Flip-flop B stays unchanged in the 0 state and $J_B = 0$, $K_B = D$ is therefore entered in the top four lines of the right-hand column. In line 5, B changes from 0 to 1 for the first time (shaded), so a $J_B = 1$, $K_B = D$ entry for the inputs is made. The reader is left to go through the rest of the table – it is easier to go right down the A bistable columns and complete the rest of the J_A and K_A inputs needed rather than go from one bistable to the other.

(e) Drawing maps for each bistable input. Four maps are needed as there are two bistables each with J and K inputs. Taking J_A first, the top line of fig. 8.10(b) shows that J_A must be 0 for $AB = 00$ and $\bar{I}+\bar{C} = 1$. $AB = 00$ maps to the left-hand column of the map in fig. 8.11 and $\bar{I}+\bar{C}$ map to the three squares shown as 0. The rest of the data in fig. 8.10(b) are put in line by line for J_A to complete the J_A map which is then examined to determine

249

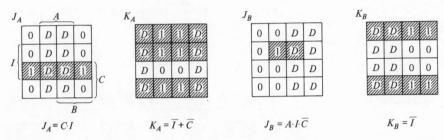

$$J_A = C \cdot I \qquad K_A = \overline{I} + \overline{C} \qquad J_B = A \cdot I \cdot \overline{C} \qquad K_B = \overline{I}$$

Fig. 8.11. Maps for the flip-flop inputs.

simple large blocks to cover the squares marked 1 and including as many D or 'don't care' states as desired. The shaded area is the best solution so $J_A = C \cdot I$ is the logic needed.

The reader is left to complete the other maps. Each map in fig. 8.11 is shaded to indicate the simplest Boolean relation to yield the four flip-flop inputs, J_A, K_A, J_B and K_B, and the appropriate expressions are written under each map.

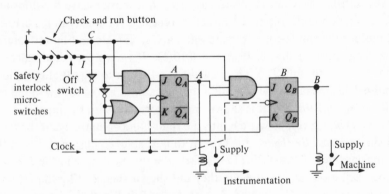

Fig. 8.12. Solution of given problem.

(*f*) *Drawing the circuit diagram.* Fig. 8.12 shows the two bistables A and B with switches on the left to yield the interlock safety signal I and check C. The J–K bistables are connected to these lines with simple gates: the reader is left to check that the gates shown do agree with the expressions determined. The outputs of the bistables can be directly connected to the instrumentation and the machine. In other cases, where the state allocation might be different, some gates may be needed here too. This is not necessarily a worse design as fewer gates may then be needed between the input lines and the bistables. Lastly, the figure shows a broken line such as would be used for clocked J–K flip-flops. These are not needed in this simple

250

sequence design. Only if this sequence were part of a larger circuit, where the timing of the events had to be synchronous, would a clock be used.

(g) *Testing the circuit.* This is an important last step – there may be eventualities at switch-on, or after a brief loss of power that have not been considered. It is important that the B flip-flop does not turn ON before A when the circuit supply is first turned on. A C–R circuit connected to a 'clear' input on B should ensure that state 0 and not state 2 would be given if the interlocks were complete at switch-on.

†8.6. Voltage level detectors, Schmitt trigger IC

To initiate most digital processes, an electrical signal which may be a resistance change, a current or a voltage change has to be converted into a low voltage to high voltage change (or vice versa) which happens quickly at some pre-set level. An example is an increasing light level which gives a rising current from a photo-detector to ring a fire-bell at a pre-set level. This first step from a simple analogue change to a fast large voltage level change is a vital stage in many digital processes.

Consider a resistive sensor, whose characteristic is shown in fig. 8.13(a), connected into the bridge which is followed by an IC-amplifier as shown in fig. 8.13(b). Suppose that the sensor has a value R_x at the critical level to be sensed. In the circuit, if R_1 is chosen to be of value R_x while $R_2 = R_3$, then the bridge circuit will be balanced so that the two inputs of the amplifier are at the same level at the critical level to be sensed. If the sensor resistance decreases, then the voltage at the non-inverting input to the amplifier, given by $R_3/(R_S + R_3)$, rises and the gain of the amplifier, which may be 50,000 or more without feedback, quickly takes the output V_o close to the supply level and the level change is signalled. If the sensor resistance increases from the level at the bridge balance state, then V_o will be close to the no-signal or earth level. A fraction of a millivolt unbalance in the bridge will be enough to effect the changeover.

Fig. 8.14(a) is a graph of the sensor output change against time and shows how the output voltage would be affected. (If a sensor of positive coefficient, and not negative as shown was used, then R_1 and R_S would have to be interchanged in the circuit.) From the graph an important defect of this simple circuit can be seen. Suppose that the value of the sensor does not change smoothly with time and there may even be noise or AC supply voltages picked up in the leads from the sensor to the circuit. This causes an *extra* output change as shown at X in fig. 8.14(a). If the number of changes in the sensor are being counted, then spurious extra

251

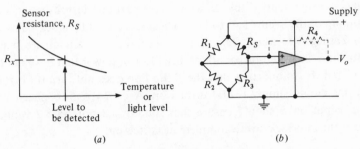

(a)   (b)

Fig. 8.13. (a) Sensor characteristic, (b) level detector circuit.

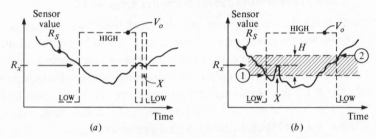

(a)   (b)

Fig. 8.14. Voltage–time waveform for level detector circuits:
(a) simple, (b) with hysteresis.

events may be registered if the level detector is used in this simple manner. The difficulty is overcome if a single high value resistor, shown broken in fig. 8.13(b), is connected between the output of the amplifier and the non-inverting input. This gives a little positive feedback. The actual value of R_4 needed to make the circuit change over when the sensor output is a few per cent *beyond* the normal changeover may be determined better by experiment than by calculation. Initially, with V_o LOW, the circuit of R_3, R_S and R_4 has to go beyond the condition for bridge balance for the change over to take place at the level (1). Later, with V_o HIGH, the change again takes place *beyond* the normal condition for changeover, now at level (2). The difference between levels (1) and (2) is shown as H or the hysteresis – or dead-band. Consider again an erratic change in the sensor value at X; with the circuit output at the HIGH state, level (2) must be reached, and *not* just level (1) exceeded, to give spurious switching. A small hysteresis band makes the circuit proof against spurious switching when the sensor changes value slowly with time while large amounts of hysteresis are useful in noisy environments.

In many cases, an IC package called a Schmitt trigger is useful as a level detector with hysteresis. It used to be made of discrete devices connected

252

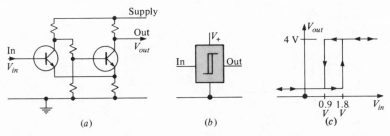

Fig. 8.15. Schmitt trigger: (*a*) circuit, (*b*) symbol, (*c*) input–output characteristic.

in the circuit of fig. 8.15(*a*) in which it is apparent that the common emitter connection of the two transistors gives feedback and hysteresis. It is now available in the common TTL range of circuits to run from a 5 V supply (7413) or in the CMOS range to run from widely varying supply voltages. The package symbol is shown in fig. 8.15(*b*). The input–output relation for the TTL device, fig. 8.15(*c*), shows that when the input voltage rises to 1.8 V the output changes to within 1 V of the supply and when it drops again below 0.9 V the output changes low again.

The main uses of a Schmitt trigger circuit are: (*a*) as a level detector to detect when some input voltage has reached a given level, (*b*) as a 'repeater', to amplify a digital sequence of low and high voltages which may be attenuated by wiring resistances (noise will be eliminated at the same time), and (*c*) to make clocks.

The last use is interesting enough to be worthy of a brief explanation. A Schmitt trigger package can be bought as a non-inverting circuit, as indicated by the input–output relation in fig. 8.15(*c*), or as an inverting circuit as shown in fig. 8.16(*a*). Here the output voltage is high until the input gets to a level V_2 when the output *drops* sharply. By using the circuit shown in fig. 8.16(*b*), a clock is made which generates abrupt voltage changes at fairly well defined periods. At switch on, the capacitor C is discharged so with input LOW, the output is HIGH. This causes C to charge up through R until at a level V_2, the switchover occurs and the output beomes LOW and close to 0 V. This causes C to discharge until the lower threshold V_1 is reached when the whole process starts again. The discharge voltage follows the well-known exponential law that $V = V_2 \exp(-t/CR)$ so the period of discharge from V_2 to V_1 is $CR \log_e V_2/V_1$. The charge-up follows a similar exponential law and the reader can prove that the charge-up period is $CR \log_e (V_3 - V_1)/(V_3 - V_2)$, where V_3 is the output of the trigger when it is high.

253

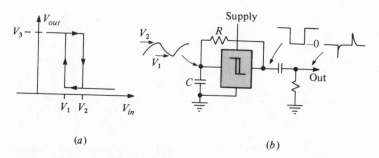

Fig. 8.16. Inverting Schmitt trigger: (*a*) characteristic, (*b*) circuit to make a simple square wave generator.

Period of clock $= CR(\log_e V_2/V_1 + \log_e (V_3 - V_1)/(V_3 - V_2))$ s. For typical values of $V_3 = 4V$, $V_2 = 2V$, $V_1 = 1$ V, the period approximates to $1.2CR$ s.

8.7 Digital to analogue converters (DACs)

These circuits convert a binary (or BCD) number in a register into a voltage or current proportional to the value of the digital number. The three most popular ways of making a converter are:

(*a*) scaled resistors into a summing IC amplifier,
(*b*) $R–2R$ resistor ladder and amplifier, and
(*c*) pulse width or pulse ratio methods.

An engineer usually has the problem of choosing the best type of converter for a particular purpose, rather than building it: so the performance of each circuit is indicated after a brief description of its mode of operation.

The scaled resistor circuit shown in fig. 8.17(*a*) is the most straightforward to understand. A digital number, say 101 ... in the register would make Q_A and Q_C HIGH and Q_B LOW. Since Q_C is connected to the summing junction of the IC amplifier through a resistor four times higher than Q_A, its weighting will be four times less. Note that for the same overall accuracy the tolerance on the larger value resistors can be less. The 10 k resistor from Q_A and the feedback resistor R_2 have to be the most accurate components in the circuit: if their tolerance is 1 per cent, there will be a ± 2 per cent tolerance at least for V_o. R_2 is chosen to scale the output to any amount desired. Suppose a level of about 8 V is wanted when all the levels of Q_A, Q_B, etc. are HIGH, and a HIGH output for the register is $+4.5$ V: then

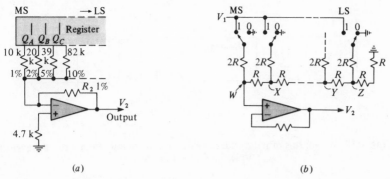

Fig. 8.17. Circuit diagrams of digital analogue converters (DACs): (*a*) scaled resistor type, (*b*) R–2R network type.

Q_A HIGH alone should give a 4 V output and, using the usual expression,

$$\text{gain of inverting amplifier} = -\frac{\text{feedback resistance}}{\text{input resistance}}$$

then

$$R_2 = 10^4 \times 4/4.5 = 8.89\text{k}$$

R_2 might be made up with an 8.2 k resistor in series with a trimmer. This circuit is not easy to build in integrated form as resistors with a range of $1:2^N$ have to be fabricated for an N-bit converter and high values are undesirable. However, for simple converters of no more than eight bits, that a user may wish to wire up say with a power IC amp, to use as a heater, its simplicity is its chief benefit. For better accuracy, the ill-defined digital output of the register would *not* feed directly into the scaled resistors but the register would drive reed or f.e.t. switches which would connect the resistors to a stable voltage source.

The R–2R network converter of fig. 8.17(*b*) allows easier IC fabrication as all the resistors can be chosen to be well matched and not too bulky. The mode of operation is as follows: consider the right-hand end of the resistor network with 2R connected to earth through the least significant switch in parallel with the two resistors in series of value R connected to earth. This is equivalent to resistance R between node Z and earth when the least significant switch is at 0; this resistance added to R gives 2R between Y and earth which is connected via the 2R path to the next input. From the symmetry of the circuit, if all the switches are closed except the most significant one, as shown in fig. 8.17(*b*), there will be two paths at the point W of resistance 2R. One is to the right and goes to earth and one up to the

255

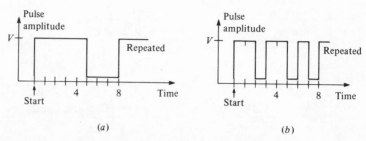

(a) (b)

Fig. 8.18. Waveforms of pulse-width convertors: (a) simple, (b) improved waveform to lessen filtering problem.

most significant switch. This will mean that W is at half the potential of the stable voltage V_1 and this is fed to the non-inverting IC circuit of unity gain and will make $V_2 = 0.5V_1$.

Consider the circuit when the most significant switch is changed over and the next most significant connects a resistor of $2R$ to V_1. The potential of X is defined by path $2R$ to V_1, another $2R$ to the right to earth, and $3R$ through W to earth. The amplifier input voltage is two-thirds of that at X and with a little analysis it is shown to be $0.25V_1$. Thus closing the switches will change the output V_2 in a way that has a binary weighting. This circuit is used widely for integrated construction of DACs.

In practice, where converters of high accuracy are needed, a combination of the two circuits described so far may be used.

The third way of converting a digital input to an analogue output is to make the input control a train of pulses of fixed frequency but with period proportional to the input count. These pulses are taken to a simple low-pass filter which will generate an output voltage proportional to the average time spent in the high state. Fig. 8.18(a) shows the pulse form that would be generated if a 3-bit binary system was converting 101 or decimal 5: the waveform would need to spend five out of its eight periods in the HIGH state. The presence of the filter causes a problem as it must have a large time constant to average the pulses and yet it must respond quickly if the digital input changes to new levels. Among the solutions to this problem is one which complicates the digital system somewhat so that the five HIGH and three LOW states of the pulse are generated as shown in fig. 8.18(b): the periods spent in the HIGH and LOW state are 'broken up' as much as possible so the filter time constant can be shorter. In accurate converters of 12 to 16 or more bits, the benefit of using the more complicated pulse generator is very great. It should be noted that no stable resistor chains are needed for this system and that it will be perfectly monotonic which

256

means that, for a smoothly rising binary input, when the most significant bit is used for the first time, the converter output will not have a change other than that expected. In converters which use scaled resistors, such as shown in fig. 8.17(a), monotonic behaviour is not assured and if the 10 k (most significant) resistor drifts a little high, the system output can be the same for a digital input of 1000 ... 00 as it is for 0111 ... 11.

The three examples given of DACs show some of the techniques that can be used. The user needs to think carefully about the stability, cost and speed of conversion that he desires and to study carefully the component data sheets for converters before making a selection.

8.8 Analogue to digital converters (ADCs)

As in the last section, three of the most used techniques will be described to show the features of these converters. They are as follows:

(a) the successive approximation method,

(b) voltage to frequency method, and

(c) dual-ramp method.

The successive approximation ADC follows the classical balancing or potentiometer method and it is shown in outline in fig. 8.19(a). At its heart is a DAC which feeds an analogue signal to a comparator which compares it to the input which is to be measured. At first, the MS (most significant) bit of the counter (which is the DAC input) is set HIGH and all the other inputs are LOW, and the comparator indicates whether the DAC output is higher or lower than the analogue input. If the DAC output is higher, then the logic sets the MS bit DAC input LOW and the next significant bit HIGH and the comparison proceeds: if the DAC output is lower, then the logic keeps the MS bit input HIGH and sets the next bit HIGH too. If the DAC is of N bits it will take N cycles to match the analogue input as well as possible. At this moment, when the LS (least significant) bit has been set the logic also 'enables' the output register to set up and to hold the binary count, which is a digital conversion of the input, while the measurement process is repeated and the output updated regularly.

Special features of the successive approximation ADC are that it is relatively accurate and fast: conversion times range from 1 to 50 μs with accuracies of eight to twelve bits commonly available.

The voltage-to-frequency ADC continuously produces a train of pulses whose frequency is proportional to the unknown input voltage. At its heart is an integrator where the input V_1 causes the integrator's capacitor to

257

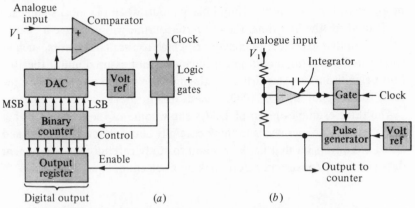

Digital output   (*a*)   (*b*)

Fig. 8.19. Circuit arrangements of analogue to digital (A to D) converters:
(*a*) successive approximation type, (*b*) voltage to frequency type.

charge. When the integrator output reaches a defined level, the gate opens
and allows the clock to trigger the pulse generator whose output is a pulse
closely defined in amplitude and length. This brief pulse sets the integrator
away from its previous level and the capacitor starts charging again. The
loop adjusts itself so that the pulse frequency gives the correct feedback to
balance the input. The method does not depend vitally on the stability of
the integrator and its capacitor as this is a balancing method. Furthermore,
drifts in the clock frequency can be made of secondary importance if it is
also divided down and used to gate the counter. If the clock runs fast, more
pulses will come out of the counter but this will be corrected by counting
them up over a reduced period.

Features of the voltage-to-frequency ADC are that the process is
continuous, unlike a successive approximation method, and the pulse
frequency follows all changes in the voltage input. The reading obtained is
the mean for the duration of the count period. If this period is made equal
to a multiple of the mains frequency, any mains interference on the input
signal will be averaged out. Also the sensitivity (the count size for a given
input) can be increased by letting the output counter run for a longer
period, which is easy to arrange. Lastly, small pulse transformers can be
used to communicate between the converter and other parts of the instrument: thus the input circuits can be floating and high voltage isolation is
readily achieved.

The third most common ADC uses the dual ramp technique in which the
unknown input voltage is used to charge a capacitor for a *fixed* time as
shown in fig. 8.20(*a*). Next, with the unknown voltage removed, a reference
voltage is used to discharge the capacitor. The time taken to discharge is a

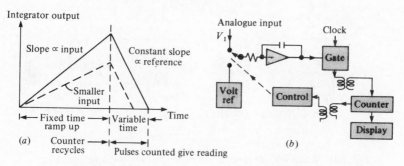

Fig. 8.20. Dual ramp ADC: (*a*) waveform, (*b*) schematic circuit.

measure of the voltage of the unknown. To ensure that the charge and discharge are linear an integrator is used. To ensure high accuracy, the end of the ramp when the voltage returns to zero must be detected carefully. A schematic circuit is shown in fig. 8.20(*b*): it is arranged that the voltage reference is of opposite polarity to the input and switching for both polarity inputs is achieved by adding a unity gain inverter to the reference so that it is switched too.

Features of the dual ramp converter are that most of the problems encountered with integrators and capacitors are reduced as it is used both in charge-up and discharge. Again, clock frequency drift is of no consequence as it defines the ramp-up time as well as being used to time the ramp down. Also, if the ramp-up time is a multiple of the mains period there is an inherent rejection of mains interference. Lastly the circuit permits very thorough isolation of the input circuits of the converter from the output and display. So the dual ramp method is used extensively in precision digital multimeters of 10-bit to 18-bit resolution where good accuracy and stability at low cost are desired but measurement speed is not important. The digital output codes are strictly monotonic with increasing input (no missing or duplicated codes as found in methods where weighted resistors might be used).

8.9 Memories; RAM, ROM, EPROM

In digital instruments, as in computers, instructions or numbers are stored in an arrangement of bistables or flip-flops, each capable of storing one bit of information. In read-and-write memory (RAM) data can be stored (or written) into the memory as well as read out. RAM originally stood for random access memory which allowed the different pieces of data to be

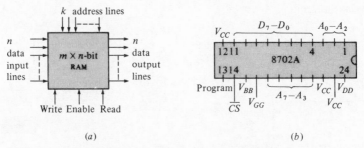

Fig. 8.21. Memories: (*a*) general $m \times n$-bit RAM, (*b*) 2k-bit ROM package.

read or rewritten at any time just by selecting the *address* of the stored bits: RAM is now also taken to mean read-and-write memory. In read-only-memory (ROM), data are permanently stored by the manufacturer or user and data can only be read later, and the stored data are not changed when the circuit is switched off. EPROM or electrically programmed read-only-memory (or erasable programmable ROM in some books) allows a user with the required equipment later to change the stored data. In use, such memories only have data read from them.

The size of a memory circuit is given in terms of the number of bits of data it can store. Since most instruments use the bits as a group or *word* of data, it is worth spending a moment on how a memory is organised. Fig. 8.21 (*a*) shows a general RAM with data input lines to the left and the 'write' input below. When the latter goes HIGH, the data on the input lines will set up n bits of the memory at the section specified by the address lines. The k address lines can specify 2^k such sections, each of n bits, so the total number of elements in the memory is $2^k \times n$ bits. For instance, a memory of 256×8-bit size would deal with 8-bit data words, it would have eight address lines as $2^8 = 256$ (suitably decoded inside the chip) and it would be called a 2K memory chip. On the right of fig. 8.21 (*a*) are the n output lines for data on which data at a specified address can be read out when the 'read' line is taken HIGH.

The enable line allows groups of memory chips to be operated together or separately. Imagine that a 16-bit word is needed in a certain circuit, then two 8-bit or four 4-bit memories could have their enable lines joined together so that the output or input lines from all the chips would give data together. The enable line can also be used as an extension of the address lines. Many microcomputers have sixteen address lines with perhaps eight of these lines joined to all the memory chips, the remaining address lines are *decoded* by separate gates whose signal is then connected to 'enable'

260

the particular memory chips desired. Memory can be realised by bipolar, MOS or CMOS fabrication methods. The relative merits of these were mentioned in § 7.11. More advanced texts will have to be consulted for more details: some are listed in appendix A. Fig. 8.21 (*b*) shows the lead arrangement of an actual memory chip, the Intel MOS 8702A 2K EPROM. There are six supply leads, eight address inputs (A_0 to A_7), eight data outputs (D_0 to D_7), a chip select pin *CS* and a program pin. As supplied, the 2048 bits are all 0. Information is stored by setting the desired word on the data lines for the address set up on the eight address lines and then applying a high current pulse to store the eight bits simultaneously. To erase the program (or memory contents), the chip, which has a transparent lid, is exposed to bright ultraviolet light for about ten minutes.

8.10   Applications

It is not within the scope of an introductory textbook to do more than hint at the extent of the applications of digital circuits. The reader should be prepared for requirements which are satisfied by a mixture of sequential and combinational techniques. Also the practical merits of compactness and good reliability often accrue from the use of just *part* of the facilities in a large-scale integrated circuit. Lastly, where the requirement is at all complex or where design changes may be needed in the future, the use of a microprocessor is ideal as it can be 'reprogrammed' as and when desired; its capabilities are described in § 8.11.

†8.10.1   Use of LSI (large scale integration) circuits as combinational logic elements

Consider the 256-bit ROM shown in fig. 8.22(*a*) and arranged as a 32×8-bit memory. One can consider it, in essence, as being an array of 32 lines horizontally (that one can address) crossing a grid of eight output lines vertically which will give an 8-bit word sequence of 1 and 0. A chip enable (*CE*) input line is provided which makes all eight outputs go high when this input is high. When the *CE* input is low, the eight output lines take up the 8-bit word that is programmed into whichever line is being addressed by the A_0 to A_4 input lines. Suppose that all of the input lines are 0, then the 0 word is addressed and as shown diagramatically a 01101010 word appears on the B_0 to B_7 output lines.

But how is this package used for logic? Imagine that we want to generate a single output which is a logical expression with five variables such as

$$\text{output, } B_0 = \bar{A} \cdot \bar{B} \cdot \bar{C} \cdot \bar{D} \cdot E + \bar{A} \cdot \bar{B} \cdot C \cdot \bar{D} + \dots$$

261

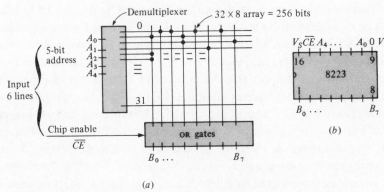

Fig. 8.22. 32 × 8-bit ROM: (*a*) circuit arrangement, (*b*) typical package.

Consider fig. 8.23 which shows part of the blank order form which a user sends away to get the memory programmed. The left-hand side of this shows how the 32 words are addressed by a binary sequence in the A_4 to A_0 input lines (with A_4 most significant). Now if we connect our five variables to the five input lines, A to A_4, B to A_3, etc., the first term $\bar{A} \cdot \bar{B} \cdot \bar{C} \cdot \bar{D} \cdot E$ will be 00001 or word 1. As we want this term to be in the output, we want a 1 specified in the *right*-hand column for the outputs, as shown. Now the next term, $\bar{A} \cdot \bar{B} \cdot C \cdot \bar{D}$ is only four variables and so does 'not care' about the condition of E, thus specifying 00100 and 00101 or the fourth and fifth input words. So we want 1 specified in two places in the B_0 column in

Customer ....
Number ....
Part Number ....8223..
Date ....

| Word | Inputs | | | | | Enable | Outputs | | | |
	A_4	A_3	A_2	A_1	A_0		B_7	B_6	$B_5 \cdot \cdot \cdot B_0$	
0	0	0	0	0	0	0				0
1	0	0	0	0	1	0				1
2	0	0	0	1	0	0				0
3	0	0	0	1	1	0				0
4	0	0	1	0	0	0				1
5	0	0	1	0	1	0				1
6	0	0	1	1	0	0				0
–										–
–										
–										–
31	1	1	1	1	1	0				
All	×	×	×	×	×	1	1	1	1 – – – 1	

Fig. 8.23. Order form for programming a 32 × 8-bit ROM.

fig. 8.23 to yield this one term. By a similar process one completes the order form for the logical expression wanted. A three-variable term requires four places in the B_0 column to be 1, a two-variable term requires eight places and so on.

It is usually not economic to order fewer than 100 identical memories and the price falls if higher production numbers are expected. The realisation in one 16-pin package of eight separate combinational signals corresponding to any combination of five input variables is most attractive.

Another such LSI package is the Programmable Logic Array (PLA). For any requirement needing ten or more gates many designers will use such circuits and much of the reduction in the chip count of recent microcomputers is the result of replacing discrete gates with PLA packages. The reader is referred to other texts for their description – see appendix A.

†8.10.2 Multiplexers and transmission gates

The digital multiplexer or selector acts very much like the selector switch S shown in fig. 8.24(a). Strictly, since the selector switch can handle analogue signals of any voltage level, its equivalent is a transmission gate, which contains basically MOSTs whose channels can be made low resistance to connect a particular signal through from input to output. Packages in the common CMOS 4000 series are called analogue transmission gates and do just this. The digital multiplexer is different in that only HIGH or LOW inputs are expected, and not the levels in between; selection of which one of several inputs is connected to the single output is determined by the 'address' on the data select lines.

Fig. 8.24(b) shows such a digital multiplexer. It has three address lines on the left which allow any one of the eight inputs to control the output. Another input marked strobe covers the uncertainty at the instant when the select addresses are being changed. When this is HIGH none of the data inputs control the output; the output is only controlled when the strobe is LOW and this inverting mode is denoted by the circle on the line to the package.

Let us consider how the multiplexer can be used to realise logic functions. We can write the output Y in terms of the data inputs D_0, D_1, D_2, etc. and of the data select inputs A, B, C as

$$Y = D_0 \cdot \overline{A} \cdot \overline{B} \cdot \overline{C} + D_1 \cdot A \cdot \overline{B} \cdot \overline{C} + D_2 \cdot \overline{A} \cdot B \cdot \overline{C} + \dots.$$

There are eight terms in this equation for Y if it is written out fully. We now use the lines A, $\dot{B}$, C for the logic input and make lines D_0 to D_7 equal to 1 where we want that particular term to be present. For example, if we

263

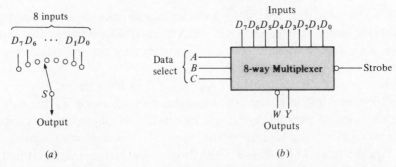

Fig. 8.24. 8-way multiplexer: (a) equivalent selector switch, (b) typical digital package.

want to make a logic function $Y = A \cdot \overline{B} \cdot \overline{C} + \overline{A} \cdot B \cdot \overline{C} + A \cdot B \cdot C$ we make data lines D_1, D_2 and D_7 all equal to 1 and the rest 0 (note A is least significant and C is most significant, so D_1 is addressed by 001 which is $\overline{C} \cdot \overline{B} \cdot A$, etc.).

However, we can do better than this because we can either make the data lines 1, or 0 or we can put a fourth variable D onto them. So the digital multiplexer can be made to realise a logic expression with one more term than the number of data select uses: in the present example a combinational circuit of four inputs and one output can be realised with the chip and a single inverter.

8.10.3 Random and pseudo-random sequences

With an N-bit shift register, it is possible to generate a repetitive sequence 2^N-1 clock periods long which is in a HIGH output state for a nearly equal time to that in which it is in the LOW state. As N gets larger, it can resemble noise and can become as a system test signal as it contains a mixture of 'high frequency' 01010 sequences and 'low frequency' 0001111100 sequences.

Fig. 8.25(a) shows the sequence obtained from a 5-bit shift register connected as shown in fig. 8.25(b). The sequence is generated by feedback through a single Exclusive-OR gate whose inputs are obtained from the last register stage (always) and one other stage. The number of this intermediate stage M for a shift register length N is:

M	2	3	3	5	6	5	7
N	3	4	5	6	7	9	10

Feedback from other stages of the register will give other sequences but they will not be of maximum length – only feedback from the $(N-M)$th stage will do equally well. The proof of this is in more advanced texts.

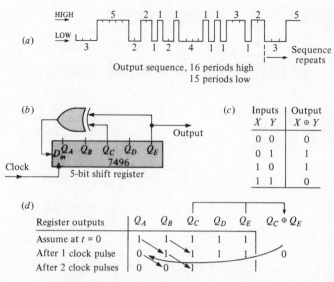

Fig. 8.25. Pseudo-random sequence generator: (*a*) waveform of 31-bit sequence, (*b*) circuit, (*c*) Exclusive-OR truth table, (*d*) truth table for generator.

Fig. 8.25(*d*) shows how the sequence is developed assuming that all the register stages are HIGH at the start. The right-hand column shows the output of the Exclusive-OR gate which is 0 in the present case with its two inputs Q_C and Q_E equal. Fig. 8.25(*c*) shows the truth table for such a gate. On receipt of a clock pulse, the 0 at data-in sets Q_A to 0 and all the other states in the register are shifted. The reader is left to develop the rest of the sequence which can be started in the way shown in fig. 8.25(*d*). The sequence is then read looking down any one column of the table to examine the successive states of one stage of the register.

It may be asked why it is a (2^N-1)-length sequence when a 2^N count is possible in an *N*-stage register. This is because a 0 in every register stage will develop a 0 as feedback and the circuit will *not* be a generator: it will be stuck with a LOW output continuously. So one condition from all those possible is missing.

8.11 Capabilities of microprocessors

The IC amplifier is at the heart of most modern analogue electronics and its manufacture revolutionised the way that measurements could be done well and cheaply. Gates, memory circuits and the digitising of data have now extended the capability of electronics immensely. A further advance which is just as great is the increasing use of microprocessors. In real cost

265

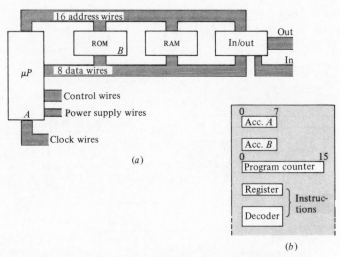

Fig. 8.26. (*a*) A simple microprocessor system, (*b*) some parts
of the microprocessor chip.

terms, they are much the same in price as the early IC designs for other
circuits. But they allow almost complete flexibility in the design of equipment. By changing one chip, a PROM which holds the program on which the
microprocessor operates, the function of the circuit can be radically
changed without any wiring changes. This is a subject for whole books:
just a few paragraphs of introduction will be given here to whet the reader's
appetite and introduce him to a few of the terms.

In fig. 8.26, a typical microprocessor chip *A* is shown connected to read-
only memory *B*, in which the program on which the system operates will
be set up and stored. It is connected by eight data wires, for an 8-bit
system, often called a bus as these wires are routed close together and
parallel to each other from chip to chip across the circuit board. Sixteen
address wires are also connected from chip to chip so that a program that
has to be drawn from the ROM step by step is achieved by changing the
address step by step; during each step the read–write control wire is set to
'read' so commanding the instructions to be put onto the data wires and
so presented to the microprocessor. The read–write memory, RAM, for
holding intermediate calculations or data, as well as the input–output chips
are also connected in the same way. Such a system is said to be *memory
mapped*. Certain address numbers are allocated to each chip, thus addresses
0 to 1000 may be allocated to ROM, 4000 to 8000 to RAM and 9000 to 9500 to
the input–output chips. There is no ambiguity then about which circuit sets
up the signals on the eight data lines.

Inside the microprocessor are accumulators which are registers on which arithmetic or logical operations can be done quickly. In fig. 8.26(*b*) two such accumulators are shown and commands such as those below may be in the instruction set for a given chip.

$$
\left.
\begin{array}{l}
\text{LOAD } A \text{ (with a number)} \\
\text{LOAD } B \text{ (with a number)} \\
\text{STORE } A \text{ (into a memory location)} \\
\text{STORE } B \text{ (into a memory location)}
\end{array}
\right\} \text{ long instructions}
$$

$$
\left.
\begin{array}{l}
\text{ADD } B \text{ to } A \\
\text{SUBTRACT } B \text{ from } A \\
\text{MULTIPLY } B \text{ to } A \\
\text{COMPARE } B \text{ to } A \\
\text{PUT } A \text{ AND } B \text{ into } A \\
\text{PUT } A \text{ OR } B \text{ into } A \\
\text{INCREMENT } A \\
\text{CLEAR } A \text{ to ZERO}
\end{array}
\right\} \text{ short instructions}
$$

The first four instructions are longer than those below because each 8-bit command is followed in the program by the memory address that a number has to be drawn from or by the number itself if it is a constant that can be specified in the program. The later instructions are complete in themselves and just eight bits long: they perform actions on the data that are assumed to be in the accumulator. These actions may typically only take a microsecond or two to complete, whereas the long instructions are slower. There, several pieces of 8-bit data have to be read to complete the memory instruction, and the action commanded may then involve several steps; even so, the longer operations may only take about ten microseconds. Thus one can do a lot of calculations, comparison of data, inputting of data and outputting of results in just one second. This is the great merit of any computer: it is quick although inflexibly bound to the program it has to follow. Writing microprocessor programs starts with the tedious step of having to observe (and learn) the instruction set of the processor being used. Then, by breaking down a problem into the most basic steps and using a flow chart, program design should involve designing for all expected eventualities. This approach will yield 'user friendly' systems which should be the aim of all engineers designing with microprocessors.

Engineers of the future can expect to need skills in programming as well as in wiring up and testing the equipment that they make.

8.12 Summary

The basic bistable gate or flip-flop yields electronic memory which is applied usefully in counters, registers, and various digital converters. Many circuits which are widely needed are available as LSI chips and, in particular, the microprocessor is revolutionising electronic design and will replace in the future most logic systems which use more than ten or so discrete circuits.

The design of sequencers is aided by state diagrams and the design of microprocessor programs is aided by the flow chart. There is no short cut to developing skills in these fields; practice and good support facilities, practical experience and discussions with colleagues and teachers is essential.

8.13 Problems

1. (*a*) Explain briefly the principle of operation of the ripple counter. Give an expression for the maximum number of available states from such a counter. State the modifications necessary to use this counter:

 (i) with a sequence of less than the maximum states used,
 (ii) in the 'count down' mode.

Identify the major disadvantage of the ripple counter.

 (*b*) Derive the characteristic equation for a *J–K* flip-flop; hence or otherwise design and sketch the logic diagram of a divide-by-5 synchronous binary counter.

(Royal Naval Engineering College: Second year)

2. A rotating shaft on a piece of machinery is to be made to provide an accurate 0 to 5 V short duration pulse once each revolution to operate an electronic counter. The shaft already carries a gear wheel with 24 teeth on it; the toothed edge of this wheel is all that is exposed.

What main components could be used to achieve the given aim? Explain briefly what type of integrated circuit packages could be used and explain any reset connections on them that would be needed.

(Cambridge University: Third year, part question)

3. The excitation table for the *J–K* flip-flop is:

$Q_n \to Q_{n+1}$		J	K
0	0	0	×
0	1	1	×
1	0	×	1
1	1	×	0

where × indicates a 'don't care' state.

What are the corresponding tables for the *T* and *D* flip-flops?

Design one decade of a synchronous, *BCD* counter using 8421 code with

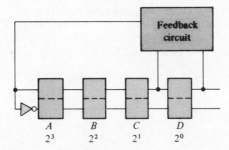

Fig. 8.27. Shift register circuit for problem 6.

J–K flip-flops and AND gates. Include a carry output on the nine to nought transition to drive further decades.

4. Make a sequential circuit with two inputs (S and Clock) that will give a logic 1 output if input S is 0 for more than ten clock pulses. The output should then stay at 1 until S is 1 again (digital alarm if no data). Assume that a MSI 4-bit binary counter package is available which has two inputs; 'count' on each negative edge and 'reset to zero' for logic 1 in. It has four outputs, A, B, C, D (least significant to most significant bit). How many NAND gates, J–K flip-flops and inverters are needed also?

5. Design and implement, using the minimum number of gates, a three-stage synchronous counter, using J–K flip-flops with the following BCD sequence:

$$4, 0, 1, 3, 2, 6, 7, 5.$$

(Royal Naval Engineering College: Second year)

6. Fig. 8.27 represents a feedback shift register using four J–K flip-flops. Feedback from C and D to A is via an Exclusive-OR gate which is constructed from four NAND elements. If the circuit is used as a Pseudo Random Number generator a sequence of fifteen numbers in a major loop should be obtained. However, with a fault in the Exclusive-OR logic the following part sequences are obtained:

(a) 8, 4, 2, 9, 4, ...,
(b) 12, 6, 11, 13, 14, ...

Suggest a possible circuit fault or faults.

(Royal Naval Engineering College: Second year)

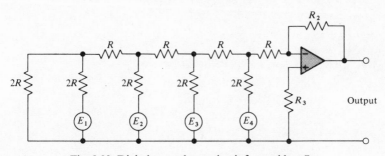

Fig. 8.28. Digital to analogue circuit for problem 7.

7. In the circuit of fig. 8.28 the voltage sources E_i ($i = 1, ..., 4$) are controlled by the digits in a 4-bit binary number. ($E_i = 5$ V for 1 and 0 V for 0). Show that the circuit acts as a DAC and determine suitable values for R, R_2 and R_3 if the output voltage range is to be 0 to -7.5 V. Ignore effects due to amplifier offset currents and voltages.

Appendix A

A list of useful textbooks

J. Millman: *Microelectronics; digital and analog circuits and systems.* McGraw-Hill, 1979.

P. Horowitz and W. Hill: *The art of electronics.* Cambridge University Press, 1980.

D. L. Schilling and C. Belove: *Electronic circuits, discrete and integrated.* McGraw-Hill, 1979.

A. B. Glaser and G. E. Subak-Sharpe: *Integrated circuit engineering.* Addison-Wesley, 1977.

Appendix B

Device data and characteristics
Type 2N3819 n-channel planar silicon field-effect transistor

* *Absolute maximum ratings at 25 °C free-air temperature (unless otherwise noted)*

Drain-gate voltage	25 V
Drain-source voltage	25 V
Reverse gate-source voltage	−25 V
Gate current	10 mA
Continuous device dissipation at (or below) 25 °C	
free-air temperature (see note 1)	200 mW
Storage temperature range	−55 °C to 150 °C
Lead temperature $\frac{1}{16}$ inch from case for 10 seconds	260 °C

* *Electrical characteristics at 25 °C free-air temperature (unless otherwise noted)*

Parameter		Test conditions			Min.	Max.	Unit
$V_{(BR)GSS}$	Gate-source breakdown voltage	$I_G = -1\,\mu A,$	$V_{DS} = 0$		−25		V
I_{GSS}	Gate cut-off current	$V_{GS} = -15$ V,	$V_{DS} = 0$			−2	nA
		$V_{GS} = -15$ V,	$V_{DS} = 0,$	$T_A = 100$ °C		−2	μA
I_{DSS}	Zero-gate-voltage drain current	$V_{DS} = 15$ V,	$V_{GS} = 0,$	see note 2	2	20	mA
V_{GS}	Gate-source voltage	$V_{DS} = 15$ V,	$I_D = 200\,\mu A$		−0.5	−7.5	V
$V_{GS(off)}$	Gate-source cut-off voltage	$V_{DS} = 15$ V,	$I_D = 2$ nA			−8	V
$|y_{fs}|$	Small-signal common-source forward transfer admittance	$V_{DS} = 15$ V,	$V_{GS} = 0,$	$f = 1$ kHz, see note 2	2000	6500	μS
$|y_{os}|$	Small-signal common-source output admittance	$V_{DS} = 15$ V,	$V_{GS} = 0,$	$f = 1$ kHz, see note 2	—	50	μS

Parameter		Test conditions	Min.	Max.	Unit
C_{iss}	Common-source short-circuit input capacitance	$V_{DS} = 15$ V, $V_{GS} = 0$, $f = 1$ MHz	—	8	pF
C_{rss}	Common-source short-circuit reverse transfer capacitance	$V_{DS} = 15$ V, $V_{GS} = 0$, $f = 1$ MHz	—	4	pF
$|y_{fs}|$	Small-signal common-source forward transfer admittance	$V_{DS} = 15$ V, $V_{GS} = 0$, $f = 100$ MHz	1600	—	μS

Notes: 1. Derate linearly to 125 °C free-air temperature at the rate of 2 mW/degC.
2. These parameters must be measured using pulse techniques. PW $\approx$ 100 ms, duty cycle $\leqslant$ 10 per cent.
* Indicates JEDEC registered data.

2N3819 *Output characteristics (measured mean of ten samples)*

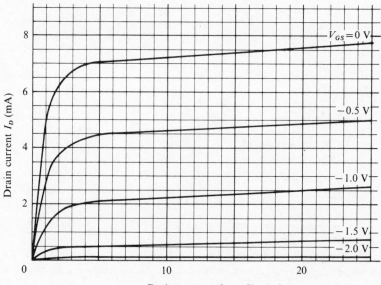

Type BC107 silicon npn planar epitaxial transistor

npn silicon planar epitaxial transistors in TO-18 encapsulation. The BC107 is primarily intended for use in audio driver stages and television signal processing circuits.

Quick reference data	
V_{CES} max.	50 V
V_{CEO} max.	45 V
I_{CM} max.	200 mA
P_{tot} max. ($T_{amb} \leq 25\,°C$)	300 mW
T_j max.	175 °C
h_{fe} ($I_C = 2\,mA$, $V_{CE} = 5\,V$, $f = 1\,kHz$)	125–500
f_T typ. ($I_C = 10\,mA$, $V_{CE} = 5\,V$)	300 MHz

Ratings

Limiting values of operation according to the absolute maximum system.

Electrical

V_{CBO} max.	50 V
V_{CES} max.	50 V
V_{CEO} max.	45 V
V_{EBO} max.	6.0 V
I_C max.	100 mA
I_{CM} max.	200 mA
$-I_{EM}$ max.	200 mA
I_{BM} max.	200 mA
P_{tot} max. ($T_{amb} \leq 25°\,C$)	300 mW

Temperature

$T_{storage}$ min.	$-65\,°C$
$T_{storage}$ max.	175 °C
$T_{junction}$ max.	175 °C

Thermal characteristics

$R_{th(j\text{-}amb)}$	0.5 deg C/mW
$R_{th(j\text{-}case)}$	0.2 deg C/mW

h-parameters (common emitter)
Measured at $I_C = 2.0\,mA$, $V_{CE} = 5.0\,V$, $f = 1.0\,kHz$

		Min.	Typ.	Max.
h_{ie}	Input impedance	1.6 kΩ	3.6 kΩ	8.5 kΩ
h_{re}	Voltage feedback ratio	—	1.8×10^{-4}	—
h_{fe}	Small-signal forward current transfer ratio	125	280	500
h_{oe}	Output admittance	—	24 μS	60 μS

BC 107 *Typical characteristics* ($T_j = 25\ °C$)

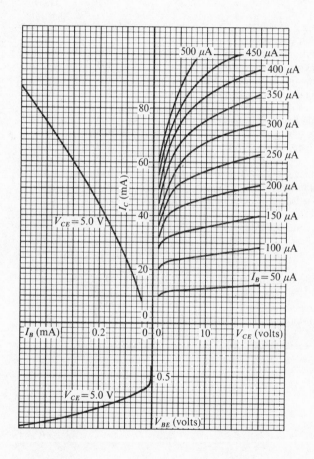

Integrated circuit operational amplifiers

Key features	μA709 General purpose	μA740 F.e.t. input High slew rate	μA741 Internally compen- sated general purpose
Input offset voltage max. (mV)	7.5	100	6.0
Input offset current max. (nA)	500		200
Input bias current max. (nA)	1500	2.0	500
Minimum voltage gain (V/mV)	15	25	20
Operating supply voltage range: min. (V)	± 9.0	± 5.0	± 5.0
max. (V)	± 18	± 22	± 18
Unity gain bandwidth (MHz)	5.0	3.0	1.0
Slew rate $A_{CL} = 1$ (V/μs)	0.3	6.0	0.5
$A_{CL} = -1$ (V/μs)	0.3	6.0	0.5
$A_{CL} = 10$ (V/μs)	3.0	6.0	0.5
Input voltage range (V)	± 10	± 15	± 15
Differential input voltage (V)	± 5.0	± 30	± 30
Temp. coefficient of input offset voltage (μV/degC)	10	20	7.0
Internal compensation		×	×
Offset adjust	×	×	×
Input protection		×	×
Output protection		×	×
Input resistance: minimum	50 kΩ	—	300 kΩ
typical	250 kΩ	10^{12} Ω	2 MΩ
Output resistance: typical	150 Ω	75 Ω	75 Ω
Common-mode rejection ratio: typical	90 dB	80 dB	90 dB
minimum	65	—	70

Data for μA741

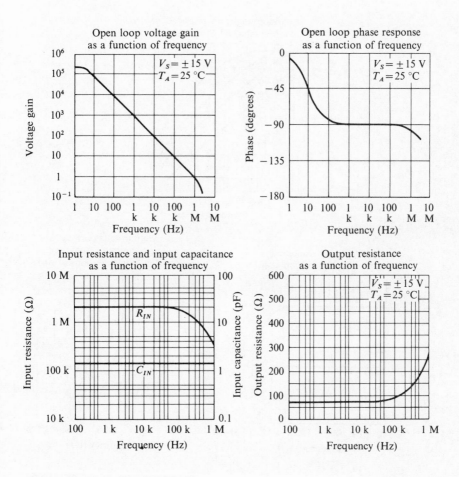

Data for μA709: *Frequency compensation curves* (*for all grades*)

Open-loop frequency response for
various values of compensation

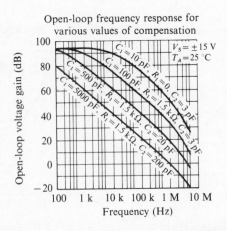

Frequency (Hz)

Frequency response for
various closed-loop gains

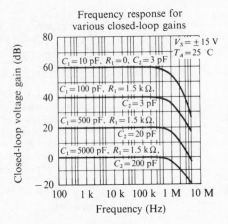

Frequency (Hz)

Output voltage swing
as a function of frequency
for various compensation networks

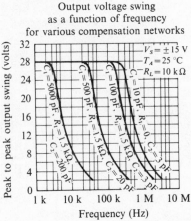

Frequency (Hz)

Protection circuits

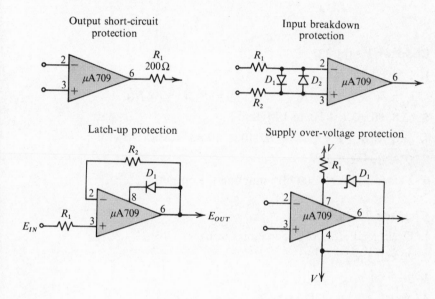

Output short-circuit protection

Input breakdown protection

Latch-up protection

Supply over-voltage protection

Answers to problems

Chapter 1   (p. 33)

1. $1\,\text{M}\Omega$, 80, $1\,\text{k}\Omega$.

2. $i_4 = IR_5/(R_4 + R_5)$, $R_5 \geqslant 99R_4$, $R_5 = R_3$, $I = Av_2/R_3$.

3. 63.8, 90, 63.1, 1 Hz to 1 MHz.

4. $I^2R_5^2R_4(R_5 + R_4)^{-2}$, $R_5 = R_4$, $10\,\Omega$, 10 mV RMS, $270\,\mu\text{F}\dagger$.

5. $C \rightarrow B \rightarrow A$, 5.2 W.

6. 2 stages, $110\,\Omega\dagger$, $300\,\mu\text{F}\dagger$ shunting the output.

Chapter 2   (p. 70)

1. Drain load $= 12\,\text{k}\Omega$, source bias resistor $= 1\,\text{k}\Omega$, -21.8.

2. $20\,\text{k}\Omega$, 3 mS.

3. 26.7, 24.3.

4. 28 dB, 318 Hz, 1.27 MHz.

5. 15.7 V, 1.25 V, 19.7 dB.

6. $\dfrac{4 \times 10^{-3}\,j\omega RR_3 C}{(1 + j\omega R_3 C)(1 + j\omega RC_1) + j\omega RC}$, 1.9 MHz, 26.6 Hz,

$$\dfrac{1}{\left(1 + j\dfrac{f}{2.5 \times 10^5}\right)\left(1 + j\dfrac{f}{3.5 \times 10^5}\right)\left(1 + j\dfrac{f}{4.5 \times 10^5}\right)}.$$

7. Drain load $= 4.7\,\text{k}\Omega$, bias resistor $= 180\,\Omega$, by-pass capacitor $\sim 500\,\mu\text{F}\dagger$.

8. $(1 + G)\,Y$, $(Y + g_0)$ (where g_0 is the output conductance of the voltage amplifier), 34.7 dB, 159.6 kHz.

Chapter 3   (p. 111)

1. $398\,\Omega$, $93\,\text{k}\Omega$, -223, $973\,\Omega$, -90, 2×10^4, -80.

2. -545.

† Nearest commonly available component.

3. -7.7, -40, 0.99 MHz, 79.6 Hz.

4. Same as problem 1.

5. 0.88, 0.80, 145 Ω, 657 Ω.

Chapter 4 (p. 158)

1. -10, 51, $R_1 = 0.91$ kΩ, $R_2 = 0.98$ kΩ.

2. $\beta_1 = \dfrac{R_p R_i}{R_1 R_i + R_p R_1 + R_p R_i}$, $\quad \beta_2 = \dfrac{R_i R_3}{(R_2 + R_3)(R_i + R_x) + R_2 R_3}$,

$R_p > 540 \, \Omega$, $R_x < 175$ kΩ, 0–3 MHz, 0–1.8 MHz.

4. 28.7.

5. $R_1 = R_2 = 1$ MΩ, $R_3 = R_4 = 18$ kΩ, $R_5 = 13.5$ kΩ, 12.1.

6. $R_E = 14.4$ kΩ, $R_2 = 1.76$ kΩ, $R_3 = 1.9$ kΩ, $v_o = 640v_2 - 630v_1$.

Chapter 5 (p. 187)

1. -9.9, 50.5 MΩ.

2. $(R + A\gamma)/(1 + A\beta)$, γ/β.

3. 100.7, 16 kHz, 1.6 MHz.

4. 50, 0.08.

5. $10 \text{ k}\Omega + 11 R_2 \leqslant 0.01 R_4$.

6. (a) -0.0096, 100 to 103. (b) -0.08, 100 to 123.

7. (a) -0.004, 2 per cent. (b) -0.008.

Chapter 6 (p. 209)

1. $1/2\pi CR$, 3180 Hz.

2. 1 kΩ, 16 kΩ, 1 μF.

3. 5010 pF, 10 pF.

4. 9, $(\sqrt{5})/4\pi T$ Hz.

Chapter 7 (p. 236)

1. (a) $X = A \cdot B \cdot C + A \cdot B \cdot D + A \cdot C \cdot D + B \cdot C \cdot D$, four 3-input AND gates + one 4-input OR gate.

(b) Five NAND gates (four 3-input, one 4-input), seven NOR gates (six 2-input, one 6-input).

2. (a) $Z = (\overline{A \cdot B \cdot C + \overline{A}}) + B$; simplifies to $A + B$, one gate.

(b) $D = A \cdot (B \cdot \overline{C} + \overline{B} \cdot C + D) = A \cdot (\overline{(\overline{B \cdot \overline{C}}) \cdot (\overline{\overline{B} \cdot C}) \cdot \overline{D}})$ = output HIGH.

3. (*a*) $Z = A + \overline{B} \cdot C$.

  (*b*) $F = \overline{A} \cdot \overline{D} + \overline{B} \cdot \overline{C} + \overline{A} \cdot \overline{C}$,

  $\overline{F} = A \cdot B + C \cdot D$.

4. $F = B \cdot \overline{C} + B \cdot \overline{D}$, $\overline{F} = \overline{B} + C \cdot D$.

5. $F = A \cdot B + A \cdot D + B \cdot C \cdot D$.

6. (*a*) $X = \overline{Y}$; (*c*) hazard on changeover of C if $\overline{A} \cdot \overline{B} \cdot D = 1$, add $\overline{A} \cdot \overline{B} \cdot D$ to Y.

7. $M = \overline{A} \cdot \overline{B} \cdot (\overline{C} + \overline{D}) + \overline{C} \cdot \overline{D} \cdot (\overline{A} + \overline{B})$, three packages, $T = \overline{H} \cdot \overline{M}$.

8. $f_1 = \overline{A} \cdot B + A \cdot \overline{B}$, $f_2 = f_1 + C \cdot B + \overline{C} \cdot \overline{A}$, one package.

Chapter 8 (p. 268)

1. (*b*) $J_A = K_A = \overline{Q}_C$; $J_B = K_B = Q_A$; $J_C = Q_B \cdot Q_A$; $K_C = Q_C$.

2. 5-bit counter with reset to middle stage to count 8 short.

3. $J_1 = K_1 = 1$; $J_2 = Q_1 \cdot Q_4$; $J_3 = Q_1 \cdot Q_2$; $J_4 = Q_1 \cdot Q_2 \cdot Q_3$, $K_2 = Q_1$, $K_3 = Q_1 \cdot Q_2$, $K_4 = Q_1$.

Carry $= Q_1 \cdot Q_4$ (normally also gated with clock).

4. Inverter, flip-flop, 2-input NAND; one of each.

5. $J_A = B \cdot \overline{C}$, $K_A = \overline{B}$; $J_B = C$, $K_B = \overline{B} + A \cdot C$; $J_C = \overline{A} \cdot \overline{B} + A \cdot B$, $K_B = A \cdot \overline{B} + B \cdot \overline{A}$.

6. Feedback only $\overline{C}$, feedback $C + D$.

7. R say 10k, $R_2 = 3.2R$, $R_3 = 0.76R$.

Quiz 1 (p. 74)

1. (*b*). **2.** (*b*). **3.** (*b*).

4. (*b*), (*a*) is also correct but not because of R–C coupling.

5. (*a*). **6.** (*b*). **7.** (*a*). **8.** (*c*). **9.** (*a*).

10. (*a*). **11.** (*c*). **12.** (*a*). **13.** (*a*), (*b*). **14.** (*d*).

Quiz 2 (p. 162)

1. (*b*). **2.** (*b*), (*e*). **3.** (*b*), (*c*). **4.** (*b*). **5.** (*a*), (*c*).

6. (*c*); (*e*) and (*f*) are older units.

7. (*a*), possibly (*b*), (*c*).

8. (*c*) **9.** (*b*). **10.** (*b*), (*d*). **11.** (*a*).

Answers to problems

Quiz 3   (p. 212)
1. (*b*), (*d*).   **2.** (*d*).   **3.** (*c*).   **4.** (*a*).   **5.** (*b*).   **6.** (*a*).
7. (*c*).   **8.** (*a*), (*b*), (*d*).   **9.** (*b*), (*d*).   **10.** (*c*).   **11.** (*b*).
12. (*c*), (*e*).   **13.** (*b*).

Index